Victorian Pantomime

Also by Jim Davis

JOHN LISTON COMEDIAN

LIVES OF THE SHAKESPEAREAN ACTORS, PART II: Edmund Kean (*editor*)

PLAYS BY H. J. BYRON (*editor*)

REFLECTING THE AUDIENCE: London Theatregoing, 1840–1880
(*with Victor Emeljanow*)

THE BRITANNIA DIARIES: Selections From the Diaries of Frederick C. Wilton
1863–75 (*editor*)

Victorian Pantomime

A Collection of Critical Essays

Edited by

Jim Davis

palgrave
macmillan

First published 2010 by
PALGRAVE MACMILLAN

Palgrave Macmillan in the UK is an imprint of Macmillan Publishers Limited, registered in England, company number 785998, of Houndmills, Basingstoke, Hampshire RG21 6XS.

Palgrave Macmillan in the US is a division of St Martin's Press LLC, 175 Fifth Avenue, New York, NY 10010.

Palgrave Macmillan is the global academic imprint of the above companies and has companies and representatives throughout the world.

Palgrave® and Macmillan® are registered trademarks in the United States, the United Kingdom, Europe and other countries.

ISBN 978–0–230–22159–8 hardback

This book is printed on paper suitable for recycling and made from fully managed and sustained forest sources. Logging, pulping and manufacturing processes are expected to conform to the environmental regulations of the country of origin.

A catalogue record for this book is available from the British Library.

Library of Congress Cataloging-in-Publication Data

Victorian pantomime: a collection of critical essays/edited by Jim Davis.
p. cm.
Includes bibliographical references and index.
ISBN 978–0–230–22159–8
1. Pantomime—Great Britain—History—19th century. I. Davis, Jim, 1949–
PN1987.G7V53 2010
792.3'8094109034222

2010023755

10 9 8 7 6 5 4 3 2 1
19 18 17 16 15 14 13 12 11 10

Printed and bound in Great Britain by
CPI Antony Rowe, Chippenham and Eastbourne

Contents

List of Table, Figures and Map

Table

Figures

Map

Acknowledgements

Many of the chapters in this book commenced life at a one-day conference on Victorian Pantomime at the Warwick Arts Centre funded by the Humanities Research Centre of the University of Warwick. I am indebted to everyone who participated in the conference for contributing to discussions that helped to give momentum to this collection of essays. I am also grateful to Paula Kennedy, Benjamin Doyle and Steven Hall at Palgrave Macmillan for their support in bringing this book into existence.

Jim Davis

Notes on Contributors

Jacky Bratton is Professor of Theatre and Cultural History at Royal Holloway, University of London. Her most recent books are *New Readings in Theatre History* (2003) and *The Victorian Clown* (with Ann Featherstone, 2006). She has recently completed a project beginning the re-cataloguing of the Lord Chamberlain's collection of nineteenth-century plays held at the British Library, and the web publication of some of the treasures found there. She is engaged on a large-scale project, in collaboration with Dr Ann Featherstone, which aims to produce a series of books and articles rewriting mid-Victorian entertainment history on a more inclusive basis. Her book on the early years of London's West End is forthcoming in 2011.

Jim Davis is Professor of Theatre Studies at the University of Warwick. He has published widely on nineteenth-century British theatre and his books include *John Liston Comedian* (1985), *The Britannia Diaries* (1992), *Reflecting the Audience: London Theatregoing 1840–1880* (2001 with Victor Emeljanow), winner of the Theatre Book Prize in 2002, and the volume on Edmund Kean in the *Lives of the Shakespearean Actors* series (2009). He is currently completing a study of the visual representation of English comic actors from 1780 to 1830.

Ann Featherstone is a performance historian in the Department of Drama, University of Manchester. Her research focuses upon popular entertainment in the nineteenth and early twentieth centuries, particularly music hall, circus, fairgrounds, penny gaffs, and how they were described in the press, journals and literature. As well as academic publications, her first novel, *Walking in Pimlico* (2009), is set in Victorian concert halls and circuses, and her second, *The Newgate Jig* (forthcoming), has the background of popular theatre and freak shows.

Richard Foulkes is Emeritus Professor of Theatre History, University of Leicester. His publications include: *Church and Stage in Victorian England* (1997), *Performing Shakespeare in the Age of Empire* (2002) and *Lewis Carroll and the Victorian Stage: Theatricals in A Quiet Life* (2005). He has recently edited a collection of essays reappraising Henry Irving, and his volume on Macready for the *Lives of the Shakespearean Actors* series was published in 2010. He was an Associate Editor for the *Oxford*

Dictionary of National Biography and General Editor of Publications for the Society for Theatre Research.

David Mayer, Emeritus Professor of Drama and Research Professor at the University of Manchester, has published extensively on British and American popular entertainment of the nineteenth and early twentieth centuries – in particular on the topics of melodrama and pantomime. His recent writings explore the interstices between the late-Victorian stage and early motion pictures. He is founder-director of The Victorian and Edwardian Stage on Film Project at the University of Manchester and was a contributing member to The Griffith Project. His books include *Harlequin in his Element: English Pantomime, 1806–1836* (1968), *Henry Irving and 'The Bells'* (1984), *Playing Out the Empire: Ben-Hur and other Toga-Plays and Films* (1994), and *Stage-struck Filmmaker: D. W. Griffith and the American Theatre* (2009).

Janice Norwood, a Lecturer in English Literature at the University of Hertfordshire, has research interests in nineteenth-century theatrical performance, especially that of the theatres of the East End of London. She has published work on the Britannia Theatre, the dramatist C. H. Hazlewood, stage adaptations of Wilkie Collins's novels, the prize fighter Tom Sayers and nineteenth-century productions of Shakespeare. She is currently working on the actress and manager Madame Vestris.

Caroline Radcliffe trained in music and acting, working freelance before completing her Ph.D. on Victorian popular theatre at Royal Holloway, University of London. She is a lecturer in Drama and Theatre Arts at the University of Birmingham.

Jeffrey Richards is Professor of Cultural History at Lancaster University. He is the author of *Sir Henry Irving: a Victorian Actor and His World* (2005) and *The Ancient World on the Victorian and Edwardian Stage* (2009), and co-author (with Katherine Newey, 2010) of *John Ruskin and the Victorian Theatre*. He is currently engaged with Professor Kate Newey on an AHRC research project on the social and cultural history of Victorian pantomime.

Jo Robinson is Head of Drama in the School of English Studies, University of Nottingham, where she teaches nineteenth- and twentieth-century theatre. She was the principal investigator for the three-year AHRC-funded project, 'Mapping Performance Culture: Nottingham 1857–67', completed in September 2009, in relation to which she is researching issues of performance and spectatorship and developing interdisciplinary research on mapping and historiographical representation. The project website is at www.nottingham.ac.uk/mapmoment.

Jill A. Sullivan is an Honorary Fellow of the University of Exeter. Her research interests lie within the field of nineteenth-century theatre and popular entertainments, with a particular interest in the development and expression of regional identity through theatre and other visual and performance genres. She is currently completing *The Politics of the Pantomime: Regional Identity in the Theatre 1860–1900* (forthcoming in 2011).

Millie Taylor began her career as a professional musical director and for almost twenty years toured Britain and Europe with a variety of musicals, and worked for several of the largest pantomime producers in the UK. She is now at the University of Winchester. Recent publications include *British Pantomime Performance* (2007), 'Integration and Distance in Stephen Sondheim's *Sweeney Todd*', *Contemporary Theatre Review*, 19/1 (2009), and 'Experiencing Live Musical Theatre Performance: *La Cage Aux Folles* and *Priscilla, Queen of the Desert*', *Popular Entertainment Studies*, forthcoming 2010. She is currently working on a book on musical theatre, *Musical Theatre, Realism and Entertainment.*

Sharon Aronofsky Weltman is Professor of English and Director of English Graduate Studies at Louisiana State University in Baton Rouge. She is author of *Performing the Victorian: John Ruskin and Identity in Theater, Science, and Education* (2007) and *Ruskin's Mythic Queen: Gender Subversion in Victorian Culture* (a *Choice* Outstanding Academic book in 1999). She also has essays either in print or forthcoming on the musical plays *Sweeney Todd, The King and I,* and *Goblin Market* and their source texts. Her current project is 'Victorians on Broadway', a book manuscript that examines the cultural work accomplished by mid to late twentieth-century musical theatre adaptations of Victorian materials.

Introduction: Victorian Pantomime

Jim Davis

In 1897 Bernard Shaw claimed that pantomime 'abuse[s] the Christmas toleration of dullness, senselessness, vulgarity, and extravagance to a degree utterly incredible to people who have never been inside a theatre', creating something that is 'a glittering, noisy void, horribly wearisome and enervating, like all performances which worry the physical senses without any recreative appeal to the emotions and through them to the intellect.'[1] Shaw's dismissal of pantomime perhaps helps to explain at least one of the reasons why such little interest has been shown, subsequently, in Victorian pantomime. Yet throughout the nineteenth century pantomime was one of the most successful and commercially viable forms of popular entertainment and a crucial component of Victorian popular culture. Moreover, pantomime as we know it today, in both form and structure, is very much the legacy of developments that occurred in the late-Victorian period.

At the beginning of the Victorian period pantomime was dominated by the Harlequinade, which featured the characters of Clown, Pantaloon, Harlequin and Columbine. The Harlequinade was preceded by an Opening, often based on legend, classical myth, nursery tales, literary classics and sometimes entirely original in content, in the course of which several of the principal characters, usually wearing what were known as 'big heads', were transformed into the Harlequinade characters. The Harlequinade then featured a pursuit of Harlequin and Columbine by Clown and Pantaloon, usually incorporating a number of comic scenes focused on Clown, who (since Grimaldi) had become the genre's major attraction. Yet, even as Queen Victoria came to the throne, change was in the air. At the Olympic Theatre in the 1830s, and subsequently at the Lyceum and Covent Garden Theatres, J. R. Planché and Eliza Vestris presented a series of 'extravaganzas' at Christmas and

Easter, drawing on classical myths and legends and on fairy tales, written in elegant rhyming couplets and usually staged with spectacular transformation scenes. Vestris herself gave impetus to the development of the modern principal boy through her lively rendition of male roles in these entertainments. In the meantime burlesque also became a popular theatrical form, drawing on Shakespeare, melodrama and many of the stories that now form the staple for modern pantomime. Through writers such as F. C. Burnand and H. J. Byron, not to mention some highly skilled performers, burlesque (with its absurdities and insatiable appetite for punning) became a popular mid-Victorian form of entertainment. It was in burlesque versions of *Aladdin* and *Cinderella* that H. J. Byron first introduced the characters of Widow Twankay [*sic*], Buttoni (Buttons) and Dandini.[2] Before long pantomime had become a hybrid form, absorbing the influence of extravaganza and burlesque (Planché, Byron, and Burnand all wrote pantomime scripts as well), and the Harlequinade began to diminish in importance. By the end of the nineteenth century, following the introduction of music hall stars and their routines and ever-increasing spectacle, pantomime had evolved once again, but continued to appeal to a wide cross-section of the community.

Pantomime in the Victorian era was not only an all-pervasive form of popular entertainment, but also functioned as a way of seeing, even as metaphor, in shaping perceptions of the contemporary world in just as forceful a way as has long been credited to melodrama. As Elaine Hadley suggests, 'a version of the "melodramatic" seems to have served as a behavioural and expressive model for several generations of English people'.[3] According to Hadley the melodramatic mode 'emerged in the early and mid nineteenth century as a polemical response to the social, economic and epistemological changes that characterized the consolidation of market society in the nineteenth century' and 'as a reactionary rejoinder to social change, but not, it must be stressed a politically reactionary response ... '[4] Hadley also refers to the melodramatic mode's crucial and significant use of non-linguistic forms of representation: 'physical gestures, political actions and visual cues', but only briefly refers to pantomime in a discussion of its possible influence on the physical and visual aspects of early melodrama.[5] Yet can we actually do justice to the impact of theatre and theatricality in shaping Victorian consciousness if we ignore pantomime and focus too specifically on melodrama? In a period when such authors as John Ruskin, Lewis Carroll, Wilkie Collins and Charles Dickens were regular visitors to the pantomime, when the pages of illustrated periodicals prior to Christmas

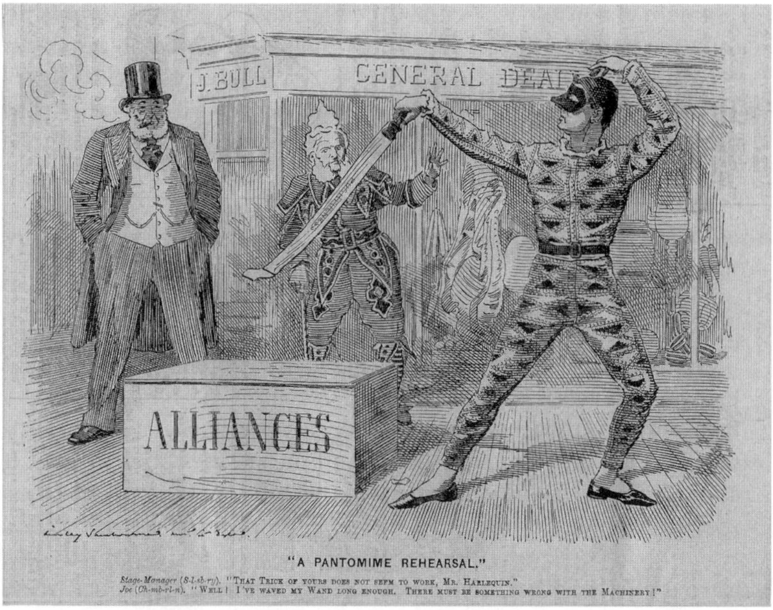

Figure 1 Pantomime as political satire: Linley Sambourne, 'A Pantomime Rehearsal', *Punch*, 24 December 1898, p. 290 (collection of Jim Davis)

regularly anticipated the proliferation of annual pantomimes, when the caricaturists of *Punch*, *Judy* and other periodicals often used pantomime images for political satire, there is surely a need to assess both its theatrical and extra-theatrical significance.

That Victorian pantomime has not been subjected to close scrutiny in the way that melodrama has, seems all the more surprising when we consider the scholarly work undertaken on Regency and eighteenth-century pantomime. David Mayer's excellent study of Regency pantomime, *Harlequin in his Element*, represents a forty-year provocation to follow up his work with a similarly insightful study of Victorian pantomime. Jane Moody and David Worrall have done further work on the social, political and material aspects and influences of pantomime in the late-eighteenth century and the Regency period. Jeffery N. Cox and Michael Garner have emphasised the importance of pantomime as a significant part of the context alongside the more literary dramas of the Romantic period. And John O'Brien, in *Harlequin Britain*, provides valuable insights into the interplay of pantomime and culture in the eighteenth century.[6]

Pantomime arguably worked as a democratic force in the eighteenth century through its cultivation of visual language. O'Brien suggests that:

> Through its deployment of dance, movement and spectacle, pantomime appealed to and in part created a new conception of the 'spectator'.[7]

Additionally, he argues that the essays of Addison and Steele in the early 1700s:

> [p]repare the way for a form of performance that could engage the attention of spectators who lacked the formal education that had traditionally been considered necessary. Pantomime's claim to be a form of universal language – an art whose silent mimeticism bypassed the distortions of speech and the vagaries of text – gave it a utopian dimension, as it expanded the possible scope of entertainment to include all orders of society.[8]

More recently David Worrall has put an emphasis on the material aspect of pantomime and its impact on consumerism, suggesting that 'playhouse audiences were perceived by manufacturers and financiers, as well as by theatre managements, to be economically active consumers and that the theatre stage was an established space where luxury products could be introduced to theatregoers'.[9] In effect pantomime is in part a materialist form, showing off commodities, encouraging consumerism and participating in the capitalist ethos of an industrial society. Worrall bases much of his discussion around *Harlequin in his Element*, but takes us into another realm altogether when he draws upon *Harlequin Negro*, a pantomime which deals with slavery and emancipation and projects the notion of Britain as 'a reforming imperial power energized by ideals of justice and sensibility'. 'What is perhaps surprising [about this piece]', says Worrall, 'is the level of political sophistication achieved through the apparently unpromising vehicle of pantomime.'[10] Jane Moody specifically follows David Mayer and, of course, Leigh Hunt, in drawing our attention to the political role of pantomime in Regency Britain and asserts that its independence from spoken language – a factor also noted by O'Brien – is what gives it this political edge. 'The virtual absence of dialogue in pantomime [Leigh Hunt] suggests, offers a form of imaginative, and by implication, satirical freedom, leaving the spectators to imagine "what supplement they please to the mute caricature before them."'[11]

By the time pantomime enters the Victorian era, then, its emphasis on spectacle and on the non-verbal allows it to function as a democratising medium socially, as a satirical and even subversive medium politically, and as a form of advocacy for consumerism materially. It may be that all these traits continued as pantomime changed its forms and structures to emerge in the more hybrid and modern manifestations that became familiar from the end of the nineteenth century. But throughout the Victorian period pantomime was perceived increasingly as family entertainment and as a significant facet of childhood experience. It also assumed a more moral tone as the century progressed and arguably lost some of the satirical sharpness that permeated Regency pantomime.

Michael R. Booth and A. E. Wilson have both sketched out key changes in the form of pantomime during the Victorian era.[12] Pantomime expanded after the deregulation of the theatres in 1843, but views on the implications of this expansion were conflicting. Disagreements over whether pantomime had improved or declined became more frequent. By 1860, according to Wilson, critics were praising improvements both in the splendour of the décor and in the literary quality of the language in pantomime. Yet there were also those throughout the Victorian period who bemoaned the decline of pantomime, most famously E. L. Blanchard, the pantomime author, and W. Davenport Adams, whose attack on new trends in pantomime in *The Theatre* of 1882 implied that the only benefit derived from pantomime was the monetary surplus that kept so many theatres open for the rest of the year. Two attributes of pantomime were singled out by Davenport Adams for particular opprobrium. The first was the infiltration of the music hall element into the genre and the damage done to youthful ears by the songs now performed:

> Such songs … would not be tolerated by pater familias in his drawing-room, and yet, when he takes his children to the pantomime, they are the most prominent portion of the entertainment. No doubt he and his children can stay away; but in that case it must be avowed that pantomime is not *virginibus puerisque*, and if it is not, what, then, is its reason for existing? It will, in that case, have to be confessed that pantomime is dead, and that burlesque of the most deteriorated type has usurped its place.[13]

Equally offensive to Davenport Adams were the cross-dressed roles and the overt display of the female form, 'a vicious object on the part of one who ostensibly provides a holiday entertainment for "the children".'[14] Surely, he argued:

> The rows of infinitesimally clothed damsels who crowd the pantomime stage are not the sort of spectacle to which it is judicious to introduce the 'young idea', especially when it is at that age at which curiosity concerning the forbidden is beginning to display itself. Over and over again must mothers have blushed (if they were able to do so) at the exhibition of female anatomy to which the 'highly respectable' pantomime has introduced their children.[15]

If Davenport Adams is to be believed, the decline of pantomime is not only structural but moral. And this is because it has become so strongly inter-linked with Victorian notions of childhood, childhood innocence and domesticity.

Pantomime's place in nostalgia for childhood and in the iconicity of the family firmly took root in the Victorian era. F. C. Burnand, for instance, recalled in detail his own childhood reminiscences of pantomime at Drury Lane:

> Oh glorious moment in the Christmas life of a town-bred child – the overture to the pantomime commences! What overture can ever equal the overture to a pantomime on the first night of performance![16]

In the 1860s E. L. Blanchard recalled being taken to the theatre as a holiday treat at Christmas and enjoying the antics of Harlequin, especially as played by Ellar, in the 1830s:

> Harlequin would then perform his feats of transformation continuously through a long series of adventures in which we were all greatly interested ... and as every trick had a sort of political or social significance a vast amount of information about passing events was concurrently imparted to the youthful spectator.[17]

For Dickens, too, pantomime is the stuff of nostalgia and childhood memories, strongly rooted in pre-Victorian times.[18] Michael R. Booth usefully suggests that, while Regency pantomime was not thought of specifically as children's entertainment, as pantomime became more exclusively family entertainment catering for children, so 'it grew simpler, less satiric, more cruel, more obvious comically and scenically, more ostentatiously moral and even instructional – in a word, less adult'.[19] Anne Varty, in her study *Children and Theatre in Victorian England*, suggests that the 'children in the pantomime audience could enter a magical world of infinite possibility, while the adults were invited to escape into a world of anarchy and subversion'. 'By appealing alike to children and "children of larger growth" ', she argues, 'age differences were similarly obscured, and the audience was constructed as an ideal body of children'.[20] Pantomime may have infantilised its audiences, in that the adult audience conceivably regressed and also entered 'a magical world of infinite possibility'. But child audiences were themselves also exposed to 'the world of anarchy and subversion', not to mention the music hall influence discussed above, intended for adults, with consequences that are briefly touched upon by Bratton and Radcliffe in their chapters in this volume.

Just as visits to the pantomime were increasingly associated with memories of childhood, so these visits provided an excuse for representing the Victorian family, at all social levels, as a unified and cohesive group, within a predetermined social structure, enjoying their leisure. The illustrated newspapers, in particular, provide a proliferation of Christmas articles and illustrations exhorting the domestic values implicit not so much in pantomime itself as in pantomime as a source of family entertainment. The existence of pantomime, then, enables the celebration of the family and also, as so many of the accompanying illustrations reveal, the hierarchies of Victorian life. In particular the vagaries of the gallery audience are depicted for the delectation of middle-class periodical readers. In 1860 *The Players* (2 January 1860, p. 7) described the gallery of a minor theatre on Boxing Night, focusing specifically on a Mrs Gullins and her four blooming daughters. 'Whistle after whistle resounds from either side of the gallery, at which sounds the head of Mrs Gullins turns to and fro, evidently hoping at length to see some face which she knows perfectly well.' On informing her daughters that a Mr Jim Bilkes will shortly join them, her eldest daughter, Miss Sarah Jane Gullins, increases [her] smile

> ... into a broad grin, at which juncture Mrs G., thinking the exposure of teeth to signify hunger, hands her a small piece of fried fish and bread from a huge basket which she holds in her lap, at the same time taking from the aforesaid basket a ginger-beer bottle, which contains liquor fit for the gods, if we may judge from the exquisite relish Mrs G. feels for it. A disturbance now takes place which is occasioned by the arrival of Mr J. Bilkes, who, seeing the object of his affection seated in the front row, shouts 'Room for one?' An answer

> in the affirmative causes him to precipitate himself helter-skelter, pell-mell, over the heads of several small boys. In his descent he upsets everybody and does severe damage to the optics of an elderly female with the heel of his boot. This leads to a disturbance, the blame of course falling on the wrong party; this leads to a general row, known to the habitués of theatres as a gallery fight ... No sooner is the curtain down [after the melodrama], than we observe Mrs G. has taken off her bonnet; her hair falls loosely, if not gracefully, over her shoulders, and she is indulging in another deep libation from that – bah! – nasty ginger-beer bottle ...

This all sounds very jovial, but the tone is condescending, containing, even voyeuristic, a repeated feature of such descriptions.

Many references to the rowdiness of pantomime audiences, particularly galleryites on Boxing Day, recur throughout the Victorian period. As I have argued elsewhere, continual accounts of unruly inebriated crowds tamed by the pantomime overture (a medley of popular songs with the gallery audience joining in) are legion, although it has to be said that sometimes the Boxing Night audiences were not always compliant. Jimmy Glover, conductor at Drury Lane, recalls the late 1860s when 'a huge eightpenny pit, and a seething mass of sixpenny and fourpenny galleryites ... catcalled the Opening on Boxing Night till not a word was heard' and shouted the latest music hall songs.[21] F. C. Wilton, stage manager at the Britannia Theatre, also recalls occasions when not a word of the pantomime was heard.[22] Generally, though, whether through the agency of the audience or through the power of performance, the pantomime created a sort of commonwealth in which the crowd came to order and willingly formed a cohesive community of boisterous but engaged spectators.

Family values, childhood, social hierarchy, orderliness derived from chaos are factors intrinsically linked to the experience of Victorian pantomime. Another question that hangs over the genre during this period is, of course, the extent to which it maintained the tradition of dissent, of social and political satire, of democratic space associated with eighteenth- and early nineteenth-century pantomime. Arguably, pantomime as a genre lost many of these qualities during the Victorian period, so that its radical potential was subsumed in topicality, increasingly spectacular displays, moral instruction (often framed through Manichean struggles between personifications of abstract values) and even vulgarity. By the end of the Victorian period pantomime had become a much more conservative affair (despite the hybridity of its form and structure) and arguably endorsed nationalism, patriotism, imperialism, the status quo, social hierarchy, racism, gender stereotyping and much else.

If pre-Victorian pantomime contributed to hegemonic shifts in cultural and social understanding, late Victorian pantomime arguably endorsed current ideologies in a much more unquestioning way. Topicality remained a feature of pantomime, but references to current events or to the local and particular were not necessarily intended as more than gently satirical, if they were intended to be satirical at all. At Drury Lane the concluding spectacles of the pantomimes, often emphasising (as previously stated) British history or British imperial might, displaced the Harlequinade. In effect the decline of the Harlequinade, a structural change to pantomime bemoaned by so many traditionalists, is indicative of the shift from pantomime as a radical spectacle to pantomime as a more conservative force. While the Harlequinade remained a strong component of the Britannia pantomime (and perhaps contributed to the continuing subversive force of pantomime in the neighbourhood of Hoxton), elsewhere it dwindled into obscurity and finally disappeared. Nevertheless, when pantomime is used as a tool for satire or a metaphor for social or political events in the Victorian era, in the work of cartoonists and caricaturists for example, it is through the Harlequinade that the points are made.

The spectacle of Victorian pantomime has already been alluded to. Almost annually articles appeared in the illustrated newspapers on preparation for the pantomimes.[23] While many hundreds of children and supernumeraries were employed on the stage itself, behind the scenes armies of scene-painters, property makers, carpenters, costumiers, hairdressers, gasmen and others were employed to make the performance possible. Despite the cost of employing so many extra bodies and the loss of revenue caused by the closure of the theatre for up to a week immediately prior to Christmas, to enable final preparations and rehearsals, pantomime generally made a profit. Mark Lemon provides a perspective on a rehearsal for a Drury Lane pantomime in a short story entitled, 'An Actor's Holiday':

> The stage of Drury Lane Theatre was in a state of 'most admired disorder.' Carpenters were at work in groups, whilst others were hauling up or lowering down pieces of scenery to and from the painting-room above. Here and there traps in the stage were open, and the master carpenter carefully superintending their 'working,' and on which depended the successful 'transformation' of the tricks which were to astonish the holiday folk. On one side of the stage a file of

Figure 2 Alfred Thompson, 'A Peep Behind the Scenes on Boxing Night', *London Society*, vol. XII (The Christmas number for 1867), p. 27 (collection of Jim Davis)

> seedy-looking men of all ages were being drilled into some extravagant actions which only acquired their full meaning when the 'property man' and the 'wardrobe keeper' had clothed them with grotesque heads and 'demon' habiliments. On the other side a number of girls in their street attire and muddy boots were simulating the graceful attitude of 'fairies'; but wanting the accessories of wings and wands and gauze-bespangled skirts, were sadly of 'the earth, earthy.'[24]

The everyday reality behind the glitter and glamour of pantomime is brought home even more forcibly by G. A. Sala, whose account of a pantomime at the Victoria Theatre in 1870, witnessed from an auditorium shrouded in fog, leads him to contemplate the ultimate mysteries of life and death.[25]

Sala's account of his visit to the Victoria pantomime reveals that he 'could not make head nor tail of its story',[26] but he confesses that this is normally the case with pantomime, a circumstance that continued until almost the end of the century, when core narratives from tales such as *Cinderella, Aladdin, Babes in the Wood, The Forty Thieves, Dick Whittington, Jack and the Beanstalk* and *Robinson Crusoe* began to dominate the genre. The absence of a clear narrative or logical plot development was a feature of mid-Victorian pantomime, noted also by Thackeray, who suggests one description will do as well as another, for '[t]he plots, you see, are a little intricate and difficult to understand in pantomime'.[27] In Sala's example:

> You are forever digging in the lucky-bag, and the bigger the bit of nonsense that comes out the better. There were two grand ballets and any number of magicians, demons, gnomes, kobballs, grim-gribbers, *loups-garous*, wehr-wolves, elfins, pixies, jack-o'-lanterns, will-o'-the-wisps, ghouls, afrits, vampires, peris, sylphides, wilis, brownies, banshees, and grotesque guards in pantomime masks. There was a Distressed Mother with a large family of children, whom she regaled with bread-and-butter, whom she put in a tub and washed, and occasionally corrected with a large birch-rod. There was singing, there was dancing, tumbling, fighting, screaming. There was plenty of red, blue, green and yellow fire; and the scenery, which was very well painted, was as gorgeous as either Dutch metal by the tone and foil-paper by the ream could make it.[28]

Among the principal London pantomimes were those at Drury Lane, the Surrey Theatre (south of the Thames), and the Grecian and

Britannia (in east London). The Grecian pantomime enjoyed particular success during the Conquest management, specifically on account of George Conquest's acrobatic skills, displayed through his extraordinary transformations into oysters, crabs, birds, monkeys, spiders and, in *Grim Goblin* (1876), even an octopus, which looked and moved just like the real thing.[29] Yet pantomime was popular not only in London, but throughout the British Isles and beyond. In Australia, for example, pantomime was performed throughout the nineteenth century, but often adapted to the interests of an Australian audience, as with Garnet Walch's *Harlequin Felix or Harlequin Laughing Jackass and the Magic Bat* (1873).[30] Sometimes pantomime aligned British and Australian interests, exercising a sort of colonial coercion. Bernard Espanasse's pantomime *Little Red Riding Hood* for the J. C. Williamson management, at Her Majesty's Theatre, Sydney, in December 1899, included the performance of a series of military evolutions by 'Our Boys', here meaning our 'girls', female dancers in the uniforms of the Naval Brigade, the Grenadier Guards, the Royal Irish, the Gordon Highlanders and the first contingent to set off to the Boer War from NSW, the New South Wales lancers. The pantomime's principal boy, Boy Blue, played by Carrie Moore in a tightly fitting khaki uniform based on that of the lancers, addressed the troops:

> My Gallant Contingent. You have volunteered for active service, and there's plenty of it to be done just now. This is your chance to show that you are *real* soldiers, not make-believe. There is plenty of fighting to be done, and you're the boys to do it. Once more we'll teach the world the lesson it has learnt before:
>
> Though Nations fall by conq'ring foes beset
> Britannia's sons were never beaten yet.
> Forward!

After successfully storming a castle of solid gold, somewhat surprisingly located in South Africa at the scene of a recent British triumph, the troops defeated its defenders (8 Boers with animal heads), rescued Red Riding Hood and raised the Union Jack. They were then informed by Red Riding Hood that they were the finest army in the world and that they always won because they never forgot that they were the Children of the Empire. Boy Blue then broke into the rousing song 'Children of the Empire, Hear Britannia call', with which the scene concluded. Immediately following this a white screen was lowered and film footage of the Boer War was shown.

* * *

This collection provides an opportunity to re-assess the significance of Victorian pantomime through a variety of lenses. Richards, Foulkes and Weltman consider what has been called the 'golden age' of Victorian pantomime through essays on E. L. Blanchard, Lewis Carroll and John Ruskin, while Norwood considers pantomime from the East End perspective. (Uniquely, the Britannia created some of the most original pantomimes of the Victorian period and maintained the traditional form of Opening and Harlequinade, successfully and effectively, for much longer than any other theatre.)

Blanchard was the most prolific author of pantomimes in the mid-Victorian period, but regretted the vulgarisation of the form as the century progressed. Ruskin and Carroll were pantomime enthusiasts, providing an extraordinary resource for insights into annual performances of the genre. The elegant, improving, moralistic pantomimes of the Blanchard era, creating the magical fairylands so admired by Ruskin, were superseded or (in Richards' words) 'defeated' by the crass commercialism and vulgarity of the Harris era at Drury Lane.

Ruskin enjoyed the pantomimes of the pre-Harris period, finding in them the means of conveying metaphorically to his readers the possibility of another world, another way of being, and the recovery of the lost innocence of childhood. Through her analysis of three *Illustrated News* images Weltman links together the aspirations of Ruskin, concerns about the employment of child performers and their contrast with the more affluent children who attended (rather than appeared in) the pantomimes. Pantomime might look back to pre-lapsarian times, but (suggests Weltman) it has the potential to be just as tarnished as the bleak and industrial world that it sought to repudiate. Lewis Carroll was very much an admirer of the Blanchard pantomimes, a number of which he witnessed with youthful companions; he also (as Foulkes reveals) befriended a number of juvenile performers. Of particular interest to Carroll were the pantomimes performed by companies of children at the Adelphi Theatre in the 1870s, an interest that some have seen as voyeuristic and even predatory, although Foulkes regards this in a more innocent light.[31] Carroll and Ruskin lost interest in pantomime as it began to change later in the century.

Nostalgia for past pantomimes, whether for the Harlequinades of the early nineteenth century or the finessed couplets and spectacular openings of the Blanchard era, is a recurrent motif in several chapters. Bratton looks at ways in which *Punch* developed perspectives on this issue in the mid-1800s, focusing in particular on a cartoon in which a small boy, still in frocks, tells his elder sister that pantomimes are not as good as they used to be. Bratton not only touches on the child audiences for whom purveyors of pantomime increasingly catered in the Victorian period, but also on cross-dressing and the 'incipient knowingness' and sexual awareness that pantomime might impart, even to its younger admirers. Aspects of these themes are further explored by Radcliffe in an analysis of Dan Leno's dame roles at Drury Lane. Leno is credited with inventing the modern pantomime dame, but also is noted for investing the role with a quasi-realistic quality. J. Hickory Wood, in his memoir of Leno, states that his pantomime roles seemed like real characters:

> His own personality was, of course, ever present; but when I saw him playing these kinds of parts, the impression he left upon my mind was not so much a picture of Dan Leno playing the part of a woman in a particular walk of life as the picture of what Dan Leno would have been if he had actually been that particular woman.[32]

Peter Holland argues that two further types of dame role were to develop subsequently to Leno. Firstly, there was the 'camp' or 'drag' type of dame and secondly the over-the-top dame, 'with ludicrously exaggerated costumes and make-up', both moving away from 'the connection between character and performer' on which Leno's dames depended.[33] Radcliffe looks specifically at the extent to which gender and sexuality might be acceptably combined in Leno's dames and at the discomfort created among contemporaries by the transformation of his Mother Goose into a younger and more glamorous version of her former self. Davis also demonstrates how representations of gender and sexuality informed the Drury Lane pantomimes, whether through Herbert Campbell's parody of the 'New Woman' or over-sexed female nursery rhyme characters, but further suggests that pantomime served the interests of those in power and helped to maintain the status quo through its depictions of race, class, militarism and imperial might.

The Drury Lane pantomimes dominated the final years of the nineteenth century, but in the opinion of many they vulgarised the finesse of pantomime as it had emerged in the mid-Victorian period. Thus our knowledge and understanding of Victorian pantomime are inevitably influenced by the likes of Ruskin and Carroll, but our most significant source for information lies in the many newspaper and journal reports published in the period. Featherstone considers the *Era's* reporting of London and provincial pantomimes in the 1870s, showing how strong the bias was towards London in these reports and how London was the touchstone for judging what went on in the provinces. Provincial pantomimes received little coverage in the national press, while the London pantomimes became the excuse for journalistic disquisitions on the genre's association with childhood and its wide appeal across the social spectrum. The provinces are, however, given their due in essays by Robinson and Sullivan. Robinson considers pantomime in Nottingham in the mid 1850sw and 1860s, examining the contexts in which it occurred and the impact of the newly-built Theatre Royal on the genre's local reception, through a process of close reading and mapping. Sullivan examines specifically the Theatre Royal Birmingham in the late nineteenth century, focusing particularly on topical allusion, an important feature of pantomime also touched on in chapters by Davis, Foulkes, Featherstone and Taylor. Sullivan concentrates exclusively on references to the local Birmingham politician, Joseph Chamberlain, examining the varied ways in which he was represented in the annual pantomimes, but also revealing the strong Liberal leanings of these pantomimes, a convention that practically became part of their audiences' 'horizons of expectation'. In the suburbs of East London, particularly at the Britannia Theatre, Hoxton, Norwood similarly reveals an expectation that the Britannia pantomimes would laud the industry and patriotism of its local working-class audiences.

Finally Taylor and Mayer explore the legacy of Victorian pantomime as manifested on the post-Victorian stage and in film. Millie Taylor, the author of a recent study of British pantomime in the twentieth century, summarises some of the developments that have taken place since the Victorian period, but shows that, as a continually evolving genre, pantomime's principal aim is to remain up-to-date and to engage with contemporary events and developments, exactly what it has always been doing. Mayer argues that, even though no pantomimes themselves were filmed, many of the gags, routines and scenic effects used in Victorian pantomime are preserved on film and points the way to future research in this area.

* * *

Victorian pantomime, according to Michael R. Booth, is the only Victorian dramatic form to have survived to the present day, despite some inevitable changes.[34] It was also one of the most common forms of popular entertainment in the English-speaking world during the Victorian period. It shaped perceptions of childhood and childhood perceptions. It encouraged nostalgia yet consistently embodied change. It both endorsed and questioned the status quo. It created disorder yet, whether in the harmony of its endings or in the attention elicited from sometimes unruly audiences, created order. It engaged with politics and social issues, and explored class, gender and race, in ways which were often paradoxical and complex, for it could be radically subversive one moment, highly coercive the next. It clearly fascinated contemporaries and arguably followed melodrama in enabling its audiences to negotiate with a wide range of issues thrown up by social and political changes. To understand the Victorians completely it is necessary to understand why this popular cultural form held such sway, not only as a commercial money-spinner for theatre managements,[35] but also as an institution in its own right, irrespective of class or location. There is much more to Victorian pantomime, as this collection hopefully reveals, than the 'glittering, noisy void' which so disgusted Shaw.[36]

Notes

1. 'Peace and Goodwill to Managers' (*Saturday Review,* 1 January 1898), *Our Theatre in the Nineties* Vol. 3 (London: Constable and Company Limited, revised and reprinted 1932, 1948), p. 279.
2. For modern editions of pantomimes, extravaganzas and burlesques by J. R. Planché, H. J. Byron and others see, Michael R. Booth, ed., *English Plays of the Nineteenth Century V. Pantomimes, Extravaganzas and Burlesques* (Oxford: Oxford University Press, 1976); H. J. Byron, *Aladdin*, ed. Gyles Brandreth (London: Davis Poynter, 1971); H. J. Byron, *Cinderella*, ed. Gyles Brandreth (London: Davis Poynter, 1971); Jim Davis, ed., *Plays by H. J. Byron* (Cambridge: Cambridge University Press, 1984); Donald Roy, ed., *Plays by J. R. Planché* (Cambridge: Cambridge University Press, 1986).
3. *Melodramatic Tactics: Theatricalized Dissent in the English Marketplace, 1800–1885* (Stanford, Calif: Stanford University Press, 1995), p. 3.
4. *Melodramatic Tactics*, p. 3.
5. Ibid., p. 4, pp. 53–6.
6. See David Mayer, *Harlequin in his Element: The English Pantomime 1806–1836* (Cambridge, Mass: Harvard University Press, 1969); Jane Moody, *Illegitimate Theatre in London, 1770–1840* (Cambridge: Cambridge University Press, 2000), pp. 208–28; David Worrall, *Harlequin Empire: Race, Ethnicity and the Drama of the Popular Enlightenment* (London: Pickering and Chatto (Publishers) Limited, 2007), *Theatric Revolution: Drama, Censorship and Romantic Period Subcultures 1773–1832* (Oxford: Oxford University Press, 2006); Jeffrey N. Cox & Michael Garner, eds., *The Broadview Anthology of Romantic Drama*

(Ontario: Broadview Press Ltd., 2003), p. xviii, 206); John O'Brien, *Harlequin Britain: Pantomime and Entertainment 1690–1760* (Baltimore and London: The Johns Hopkins University Press, 2004).

7. *Harlequin Britain: Pantomime and Entertainment*, p. xxi.
8. Ibid., p. xxii.
9. *Theatric Revolution: Drama, Censorship and Romantic Period Subcultures 1773–1832*, p. 265.
10. Ibid., p. 293.
11. *Illegitimate Theatre in London, 1770–1840*, p. 227.
12. A. E. Wilson, *Pantomime Pageant* (London: Stanley Paul & Co. Ltd., 1946), *Christmas Pantomime: The Story of a British Institution* (London: Allen & Unwin, 1934); Michael R. Booth, *English Plays of the Nineteenth Century V. Pantomimes, Extravaganzas and Burlesques* (Oxford: Oxford University Press, 1976). The introduction to this volume is reprinted in Michael R. Booth, *Prefaces to English Nineteenth-Century Theatre* (Manchester: Manchester University Press, 1980), pp. 149–211. Gerald Frow, *'Oh, Yes It Is': A History of Pantomime* (London: BBC Books, 1985) and Raymond Mander & Joe Mitchenson, *Pantomime: A Story in Pictures* (London: Peter Davies, 1973) are also useful resources on this topic.
13. W. Davenport Adams, 'The Decline of Pantomime', *The Theatre* n.s., vol. V (1 February, 1882), p. 89.
14. Ibid.
15. Ibid.
16. *Records and Reminiscences*, quoted in Wilson, *Christmas Pantomime*, p. 140.
17. Clement Scott and Cecil Howard, *The Life and Reminiscences of E. L. Blanchard* (London: Hutchinson & Co, 1891), vol. 2, p. 579.
18. Richard Findlater, ed., *Memoirs of Joseph Grimaldi* by Charles Dickens (London, 1838; revised edition London: MacGibbon & Kee Ltd., 1968), p. 11.
19. *English Plays of the Nineteenth Century V. Pantomimes, Extravaganzas and Burlesques*, pp. 42–3.
20. Anne Varty, *Children and Theatre in Victorian Britain: 'All Work, No Play'* (Basingstoke: Palgrave Macmillan, 2008), p. 147.
21. J. M. Glover, *Jimmy Glover His Book* (London: Methuen, 1911), p. 159.
22. Jim Davis, ed., *The Britannia Diaries 1863–1875* (London: The Society for Theatre Research, 1992).
23. See Booth, *English Plays of the Nineteenth Century V. Pantomimes, Extravaganzas and Burlesques*, pp. 485–518 and Russell Jackson, ed., *Victorian Theatre* (London: A & C Black, 1989), pp. 180–1, 193–200, 231–3, for more detail on the staging of Victorian pantomimes.
24. *London Society*, vol. XII (The Christmas number for 1867), p. 18.
25. G. A. Sala, 'Pantomime and Pandemonium: Two Nights in the New Cut', *Belgravia*, vol. XI (April 1870), p. 176.
26. Ibid., p. 175.
27. W. M. Thackeray, 'Round About the Christmas-Tree', *Roundabout Papers, reprinted from the* Cornhill Magazine (New York: Harper and Brothers Publishers, 1863), p. 135.
28. G. A. Sala, 'Pantomime and Pandemonium: Two Nights in the New Cut', p. 175.

29. Frances Fleetwood, *Conquest: The Story of a Theatre Family* (London: W. H. Allen, 1953), p. 91.
30. Garnet Walch, *Harlequin Felix or Harlequin Laughing Jackass and the Magic Bat*, ed. Veronica Kelly (St Lucia: University of Queensland Press, 1988); see also Margaret Williams, *Australia on the Popular Stage 1829–1929: An Historical Entertainment in Six Acts* (Melbourne: Oxford University Press, 1983), pp. 57–86.
31. See Varty, *Children and Theatre in Victorian Britain*, pp. 50–8 for an alternative take on Caroll's predilection for child performers.
32. J. Hickory Wood, *Dan Leno* (London: Methuen, 1905), p. 119.
33. Peter Holland, 'The Play of Eros: The Paradoxes of Gender in English Pantomime' *New Theatre Quarterly* 13.51 (1997), p. 201.
34. Michael R. Booth, *Theatre in the Victorian Age* (Cambridge: Cambridge University Press, 191), p. 202.
35. See Tracy C. Davis, *The Economics of the British Stage 1800–1914* (Cambridge: Cambridge University Press, 2000), pp. 313–14, pp. 342–7.
36. Currently there are two major funded research projects in existence devoted to the topic, one based at the University of Glasgow and one based at the Universities of Birmingham and Lancaster.

Part 1

The Golden Age of Pantomime: The Mid-Victorian Period

1

E. L. Blanchard and 'The Golden Age of Pantomime'

Jeffrey Richards

The distinctive Victorian pantomime emerged in the middle of the nineteenth century. Its predecessor, the Regency pantomime, the clown-centred Harlequinade, reached its apogee during the pre-eminence of Joseph Grimaldi (1806–1837). The Regency pantomime had a short one or two scene 'opening' with a plot derived from fairy story, nursery rhyme, myth and legend and the much longer Harlequinade in which the characters of the opening were transformed into the characters of the *commedia dell'arte* – Harlequin, Columbine and Pantaloon – and engaged in a knockabout sequence of song, dance and acrobatics. The largely dialogue-less form of the Regency pantomime was dictated by the 1737 Licensing Act which gave a monopoly of the spoken word on stage to the patent theatres, Drury Lane and Covent Garden.

However the passing of the Theatre Regulation Act of 1843 signalled a major shift in the nature of the genre. The act abolished the patent theatres' monopoly of the spoken word and opened up the use of dialogue to all theatres. This had a direct effect on the 'opening' which now got longer and longer, revelling in the linguistic freedom allowed by rhyming couplets, the ability to pun and the chance to comment on current events. At the same time another popular form was grafted onto the pantomime. This was the extravaganza, which burlesqued popular legends and folk tales and combined this with glittering spectacle. The form, first seen in *Olympic Revels* (1831), was largely the work of J. R. Planché (1796–1880). The previous dominance of Harlequin and Columbine was challenged by the rise of the principal boy and the dame as the characteristic figures of the pantomime. Various explanations have been advanced for this development. The most plausible is that the principal boy, a female performer in male dress, appeared following the success of Madame Vestris in the breeches roles in the Olympic and

Lyceum extravaganzas, and that the dame, a man in drag, appeared in the wake of the recruitment of music hall performers into pantomime casts. As the opening lengthened, so the Harlequinade shrank. By the 1890s the opening ran to several hours with the Harlequinade reduced to a couple of scenes. The attractions of the pantomime were by then spectacle, lavish ballet sequences, troupes of child performers, music hall favourites and high quality scene painting.

The transformation in the nature of the pantomime makes it as a genre a manifestation of the phenomenon described by Diane Purkiss: 'It is in Victorian England that fairyland ... undergoes a popular explosion. Fairies, elves, gnomes and small winged things of every kind multiply into swarms and infest writing and art and the minds of men and women.'[1] In France and Germany the fairy tale became a new art form and the expression of the new Romantic sensibility, which included a fascination with the supernatural. In England the fairy tale, discouraged by the forces of Puritanism, was revived as part of the reaction against the regimentation, depersonalization and materialism associated with the Industrial Revolution and the Factory System. New fairy tales were written in the 1840s and 1850s in particular in order to critique the failings of the urban and industrial world of the nineteenth century.[2]

Collections of fairy tales began to appear in English. In 1823 and 1826 there appeared an English translation by Edgar Taylor of Grimms' fairy tales, *German Popular Stories*, illustrated by Cruikshank. It has been called 'the most important publication to stimulate an awakened interest in fairy tales by children and adults'.[3] In 1827 Carlyle published a selection of German fairy tales in two volumes, *German Romances*. In 1840 there was a new translation by Edwin Lane of tales from the *Arabian Nights*. In 1846 the first translation of Hans Christian Andersen's fairy tales appeared. In addition new fairy stories were being written by noted writers in England: among them Dickens, Carlyle, Thackeray, Hood and Ruskin.

Where almost all the fairy tales of the 1840s and 1850s were allegorical critiques of modern industrial capitalism, the genre diverged from 1860 onwards as the majority of fairy story writers (e.g. Mrs Craik, Andrew Lang and E. Nesbit) sought to reconcile their readers to the status quo while a minority questioned the existing social structure and values (Lewis Carroll, George MacDonald, Oscar Wilde). At the same time fairies were desexualized and linked with the idea of childhood innocence. Dickens wrote an essay for *Household Words* (1 October 1853) on the ethical value of fairy stories, arguing that they taught 'forbearance, courtesy, consideration for the poor and aged, kind treatment of animals, the love of nature, abhorrence of tyranny and brute force'.[4] This became a widely accepted view.

There was a strong tradition of fairy painting in Victorian England, as exemplified by the work of Richard Dadd, Joseph Noel Paton, Dicky Doyle, John Anster Fitzgerald (known as 'Fairy Fitzgerald') and Arthur Rackham, who illustrated Barrie's *Peter Pan*, one of the later and most influential celebrations of fairies. *A Midsummer Night's Dream* inspired 135 paintings between 1750 and 1900, chiefly of the fairy scenes.[5] Some of the fairy painting was dark, sinister and disturbing, particularly the work of Dadd who went mad, murdered his father and spent the rest of his life in a lunatic asylum where he continued to paint fairyland. John Ruskin was a powerful advocate of the school of fairy painting but he preferred the more wholesome and innocent creations of Kate Greenaway and Helen Allingham.[6] It is clear that he associated fairies directly with childhood innocence and fairyland with a beautiful and spiritual pre-industrial landscape, an ideal and idealized pre-lapsarian world.

He devoted one of the six lectures that made up *The Art of England* to 'Fairyland'.[7] In it he claimed to have been brought up 'principally on fairy legends' and to entertain still a predilection for them.[8] He actually wrote a popular and much re-printed fairy story *The King of the Golden River* (1850) and he persuaded the publisher J. C. Hotten to reissue Edgar Taylor's English translation of *Grimm's Fairy Tales* (1869), contributing an introductory essay to the collection.

Despite the importance of fairy story books and the vogue for fairy paintings, the place where Fairyland came alive for most Victorians was the theatre. It was an article of faith in Victorian Britain that the fairy tale was one of the highest and purest forms of literature. This extended to its stage manifestations. Tracy Davis records: 'In addition to the Shakespearean canon ... and W. S. Gilbert (the only Victorian dramatist of fairy fare that endures in popular memory), fairies appear in productions from every decade and in every genre ... To some people, fairies were commensurate with theatre and theatrical practice, indelible to the idea of theatre as a metonym of faith in magicality.'[9] It was the pantomime which was the principal and continuing vehicle for the transmission of fairy lore in its most vivid and immediate form to an audience of both children and adults.

The intrinsic value of pantomime was confirmed by that most severe and austere of critics, William Archer, the translator and advocate of

Ibsen and the theatre of ideas, who admitting in *The Theatrical World of 1893* that 'the full glory of the mimic world did not burst upon me until I saw a Drury Lane pantomime – *Beauty and the Beast*', declared the pantomime 'a national art form – or perhaps I had better say a national institution', going on to say:

> It is, in its essence, an invaluable art-form, far superior to the French *revue* ... There is room within its infinitely flexible, expansible framework for all sorts of ingenious and delightful developments – for poetry, fantasy, parody, satire, sense, nonsense, the most ingenuous nursery babble and the most penetrating criticism of life. The ideal pantomime should charm the senses, stimulate the imagination, and satisfy the intelligence. It should be an enchanting fairy-tale to the young, to the old a witty, graceful, genially-satiric phantasmagoria. It has this immense advantage over burlesque, that it does not necessarily involve the degradation of anything noble or beautiful. The nursery folk-lore in which it finds its traditional subjects presents just the requisite blending of the graceful with the grotesque.[10]

Many notable and prominent Victorians were to be found in the audiences at the pantomime: among them John Ruskin, Charles Dickens, Lewis Carroll, Matthew Arnold and W. E. Gladstone.

The undisputed master of this theatrical form was Edward Leman Blanchard (1820–89). A typically prolific mid-Victorian man of letters, he wrote dramas, songs, guide books and theatrical criticism. But he was best known for his pantomimes. *The Illustrated London News* (1 January 1876) called him 'the prince of modern pantomime inventors'. *The World* called him 'the genius of pantomime'.[11] The comic magazine *Judy* presented him with the Ally Sloper Award of Merit as 'The King of Pantomime Writers'.[12] Modern pantomime historian Gerald Frow has written: 'He was one of the most prolific, the most literary, the most consistently inventive of pantomime writers and his work stands as a monument to the charm, grace, prettiness and true delights of Victorian pantomime.' The classic Blanchard pantomime combined a book of inventive rhyming couplets, the imaginative telling and sometimes interweaving of fairy tales, folk tales and nursery rhymes, lengthy and elaborate ballet sequences, beautiful scene painting and moral lessons. This came to be seen as the required form for the pantomime and Blanchard would become the hero of those who valued the 'traditional' pantomime, as it came to be regarded.[13]

Blanchard was a much-loved figure. Clement Scott, a lifelong friend and the co-editor of his diaries, wrote of his 'gentle, considerate and kindly nature', that he 'never lost the heartiness, the impulse, and the energy of a boy', that he was a 'sunny, light-hearted, affectionate companion', that he 'could not bear giving pain to any one', that he loved children and animals, that 'his tastes were simple and his habits healthy' and that 'friendship, nature and tobacco were the solace … of his life'. All these qualities can be seen reflected in his pantomime scripts. What Scott and his other friends did not know was that he was burdened by perpetual money worries as he sought to support single-handed 'innumerable relatives' and was continually grieving for his separation from the woman he loved, who was on the other side of the world. Eventually she returned to marry him and 'Carina gave to this lonely, dispirited, and forlorn man the happiest hours of … his life'. But the punishing schedule of work he undertook eventually took its toll on him and when he died in his sixty-ninth year in 1889, the doctor said: 'It was a case of gradual decay. He was "worn out" '.[14] Interestingly Blanchard was at one time a keen spiritualist. But his faith in spiritualism was severely dented, though not entirely extinguished, by the exposure of the Davenport Brothers as frauds and the downfall of Daniel Dunglas Home, parodied by Browning as Mr Sludge the Medium.

E. L. Blanchard provided the annual Drury Lane pantomime for nearly forty years (1852–88), under the managements of E. T. Smith (1852–62), Edmund Falconer (1862), Edmund Falconer and F. B. Chatterton (1863–66), Chatterton alone (1866–79) and finally Augustus Harris (1879–96). The combination of Blanchard as writer, William Beverley as scenic designer and a succession of managers willing to spend lavishly meant that Drury Lane set the standard for everyone else to beat. Reviewing the 1864–65 offering *Hop O' My Thumb And His Eleven Brothers*, *The Times* (27 December 1864) acknowledged the winning combination at Drury Lane, praising the Falconer–Chatterton regime for doing everything they undertake 'in the best manner that circumstances will allow' and for attaching 'great importance to their pantomime'. 'All the artists are of the best … The painter is Mr. William Beverley, the acknowledged chief of faery illustration, the genius to whom the "transformation scene" in the present sense of the word may be said to owe its existence.' The author, E. L. Blanchard, supplies 'invariably successful works' and is 'distinguished as a comic lyrical poet of no small pretensions'. He provided not only the Drury Lane 'annual' but pantomimes for many other theatres. In 1852 alone he had five pantomimes playing at different London theatres.

Year after year an encomium upon Blanchard would appear in the press. Typical is this article from *The Era* (31 December 1871):

> Age does not stale nor custom wither the infinite variety of our Christmas poet ... out of his imagination pour Fairies, Will o' the Wisps, Sprites, and all the Spirit-world of Christmas-time, but he is, indeed, the enchanter, for he possesses the singular art of never repeating himself, and, on the contrary, of making each year's Pantomime superior to the last. For twenty-two years this magician, this Father Christmas, has wreathed us in smiles and pelted us with the toys of his fancy ... This is emphatically a pantomime for the children – the children so loved by the author ... It flows over with loving-kindness and that particular delicate and pure fancy which suits the innocence of childhood, and is so esteemed by those of a larger growth who would imitate childhood's simplicity.

The moral message was particularly important to Blanchard. It provided 'the point' of the show. He was critical of pantomimes without a message, describing the 1855 pantomime *Harlequin and the Maid and Magpie* at the Princess's Theatre as having 'a badly written opening ... no point about it' and the 1862 pantomime *Lady Bird or Harlequin Lord Dundreary* at the Westminster Theatre as 'dull and pointless'.[15]

Blanchard's regular message was one of consensus and 'improvement'. His very first Drury Lane pantomime, *Harlequin Hudibras* (1852–3) saw an alliance established between Antiquity and the Spirit of Improvement, in which the best of the past would be retained along with the benefits of modernity. In *Hey Diddle Diddle or Harlequin King Nonsense and the Seven Ages of Man* (1855–56) the Spirit of Common Sense proposes a compromise with King Nonsense to allow more intellectual material into infant literature and goes on to defeat Red Tape and Routine and banish them to Noodledom. In *Little Jack Horner* (1857–58) the souls of children are competed for as Intelligence, aided by Imagination, Discovery and Invention, battles against Ignorance, backed by the spirits of Prejudice and Superstition.

Several pantomimes hinged on teaching girls the requisite qualities for womanhood, marriage and family life. *Jack and Jill, Harlequin King Mustard and the Four-and Twenty Blackbirds Baked in a Pie* (1854–55) had Jill learning the culinary skills appropriate for a wife. In *See-Saw, Margery Daw* (1856–57), the idle and bullying Margery is converted to the practice and principles of needlework. In *Jack in the Box* (1873–74) the spoiled Princess Poppet learns the virtues necessary in a wife – patience, economy, industriousness and humility.

The *Illustrated London News* (2 January 1875) endorsed the importance of the moral message when it declared:

> Might we not find piety in a pantomime and as much 'good' in a harlequinade as in 'everything else'. Every skilful concocter of a pantomime feels that he must provide a meaning for it, and that its idlest scene must have a significance. True it is the product of the wildest fancy: but then fancy is a sacred faculty. To it we are indebted for Spenser's 'Fairie Queen' and Shakespeare's 'Midsummer Night's Dream', works which symbolise spiritualism [spirituality] the most abstruse ... We may suggest that there may be more in a pantomime than meets the ear or the eye ... Of modern pantomime-writers who have thought thus worthily of pantomime-writing and carried their thoughts into act and scene, Mr E. L. Blanchard claims the highest rank. His openings are always carefully written, and his rhymed couplets emulate the elegance of Pope in 'The Rape of the Lock' or 'The Dunciad'. True, the satire is of the mildest form, and ought to be: else, wherefore is the disguise of allegory assured.

As important as the moral message was the visual appeal and here the Blanchard pantomimes were able to draw on the talents of one of the greatest scene-painters of the nineteenth century, William Roxby Beverley. Beverley (1810–1889) was the youngest son of William Roxby (1765–1842), a well-known actor-manager who performed under the name William Roxby Beverley. The younger William became a scene-painter, working for his father and brothers who controlled the Durham circuit of theatres. His first London engagement was in 1839 painting the scenes for the pantomime *Baron Munchausen*. From 1847 to 1855 he worked at the Lyceum, under the management of Madame Vestris, providing the scenes for among other productions the extravaganzas of J. R. Planché. In his autobiography, Planché, who regarded Beverley as combining the pictorial talent of the artist Clarkson Stanfield and the mechanical ingenuity of the mechanist William Bradwell, expressed some exasperation that Beverley's scene-painting came to overshadow his texts: 'Year after year Mr Beverley's powers were tasked to outdo his former out-doings. The *last* scene became the *first* in the estimation of the management. The most complicated machinery, the most costly materials, were annually put into requisition ... As to me, I was positively painted out. Nothing was considered *brilliant* but the last scene.'[16] This was because Beverley had perfected the transformation scene, the often spectacular sequence that linked the opening and the Harlequinade.

In 1854 he began his association with Drury Lane and painted the scenery both for Shakespeare and the pantomime until the demise of the Chatterton regime in 1879. Recalled by Augustus Harris for the 1883–4 pantomime *Cinderella,* he contributed one scene each year, despite failing eyesight, until the 1887–88 *Puss in Boots*. Blanchard did not share Planché's sense of irritation with the praise accorded to the scene painting. He dedicated the published script of his pantomime *Riquet With The Tuft* (1863) to 'Mr William Beverley whose artistic ability and exquisite taste in embellishment for so many years enabled the Wonders of Fairyland to be realised on the English stage, which he has done so much to adorn by his pictorial skill and poetical fancy.'

Beverley's scenes would regularly stop the show as he was called forth sometimes more than once to take a bow before the action could proceed. Typical were the two most praised scenes in *Hop O' My Thumb* (1864–5). The 'Valley of Mosses and Lichens at Daybreak' featured a series of grottoes reflected in a stream running through them, fairies reclined on rows of rocks, mosses hung in clusters overhead and sleeping sunbeams awakened to dance in yellow light as the Man in the Moon descended to enlist their aid against the Ogre Fee-Fo-Fum. The transformation scene, 'The Ascent of the Rays of the Golden Light', saw the clouds disperse to display the glories of the sun, while fairies floated up and down before the rays. This scene, said *The Times* (27 December 1864), by its transformation of colours 'gives a clearer notion of faery life passed in the air than any exhibition hitherto seen upon the stage'. Beverley was said to have 'surpassed himself' in these two tableaux. But every year Beverley earned praise for his fairy scenes and landscape painting. 'The Fairy Factory of the Wheels of Fortune' in *Little King Pippin* with its wheels of burnished gold and all the operatives, young women in shining raiment, was, said *The Times* (27 December 1865) 'a dazzling combination of brilliant colours and revolving figures … that surpasses even Mr. William Beverley's previous achievements'. And so it went on through 'The Fairy Boudoir' in *Faw, Fee, Fo, Fum or Harlequin Jack the Giant Killer* (1867–8); 'the Web of Golden Gossamers' in *Grimalkin the Great or Harlequin Puss in Boots* (1868–69); the 'Haunt of the Water Nymphs' in *The Dragon of Wantley* (1870–71) down to the final transformation scene in the final Chatterton pantomime, *Cinderella* (1878–79) 'The Assemblage of the Hours'. *The Times* (27 December 1878) called it 'a brilliant combination of colours irradiating numberless fairy forms'. For several generations of children, Beverley had brought to life the fairy world and its denizens on the stage at Drury Lane.

Blanchard was both willing and able to tailor his scripts for particular performers, and during his long writing career there were a number of changes of personnel on the stage at Drury Lane. For several years the pantomimes were built around the talents of a diminutive juvenile star, Master Percy Roselle. Roselle had stolen the show in *Harlequin Sinbad the Sailor* (1863–4), playing the Old Man of the Sea, the king of the pygmies. *The Times* (28 December 1863) proclaimed him 'a most extraordinary specimen of precocity, speaking with the utmost distinctness, and representing with terrible truth the inebriety which causes him to fall from the shoulders of his victims'. For the next four years he was the star of the pantomime. He appeared in the title role of *Hop O' My Thumb* (1864–5), *Little King Pippin* (1865–6), *Number Nip* (1866–7) and as the Cornish Puck Pigwiggin in *Faw, Fee, Fo, Fum* (1867–8). But the following year he was missing and was not seen again at Drury Lane. In his brief heyday he had 'delighted' Blanchard and enchanted Lewis Carroll who wanted to write a play as a vehicle for him.[17]

The Roselle era was succeeded by the age of the Vokes family. The Vokes family were recalled as 'distinguished and unique' among theatrical families by critic H. G. Hibbert.[18] The Vokeses were the children of theatrical costumier Frederick Vokes and one after another they made their debut in child roles on the legitimate stage. Fred Vokes (1846–88) was the star of the family. Trained by his father as a tailor, he soon abandoned this for the stage, making his debut at the age of eight. He was taught dancing by Flexmore, the clown successor of Grimaldi, and became a leading eccentric dancer. Fred was joined on the stage by his three sisters, Jessie (1851–84), who took care of the troupe's business affairs, Victoria (1853–94) and Rosina (1854–94). They first teamed up as the Vokes Family in 1861 in Edinburgh and made their pantomime debut in *Humpty Dumpty* at the Lyceum Theatre in 1868–9. But in 1869–70 they moved to Drury Lane for *Beauty and the Beast* and remained there for ten years missing only the pantomime *Jack in the Box* (1873–4) when they were on tour in America.

Victoria and Jessie never married. But in 1877 Rosina married Cecil Clay, brother of the composer Frederick Clay, and retired from the troupe, formed her own company and successfully toured America in a programme of comedies. She was replaced in the troupe by Fred's wife, née Bella Moore. The troupe had been joined by a non-family member Walter Fawdon who performed under the name Fawdon Vokes and outlived them all, dying in 1904. Under the influences of the Vokeses, dance became a predominant element in the pantomime and the fairy tale element took precedence over the comedy.

F. B. Chatterton under whose regime the 'traditional pantomime' reached its apogee, had been stage struck from an early age. He became lessee of the Theatre Royal, Drury Lane, initially in partnership with the playwright Edmund Falconer (1862–66) and subsequently alone. His great ambition was to 'restore old Drury to its position as the home of the poetic drama'[19] and to that end he staged lavish productions of both Shakespeare and Byron. But his productions of *Antony and Cleopatra, Richard III* and *The Winter's Tale* all failed spectacularly, causing Chatterton to coin the much repeated aphorism: 'Shakespeare spells ruin and Byron bankruptcy'. He also very deleteriously entered into partnership with Benjamin Webster to run the Adelphi and Princess's Theatres, an act he subsequently described as 'suicidal'[20] as losses sustained in these theatres drained the profits he made at Drury Lane on, for instance, a series of successful adaptations of Scott's Waverley novels.

He had always relied on the success of the pantomime to bale him out but in 1879 this recourse failed. Chatterton laid the blame squarely on the Vokeses, telling actor-manager John Coleman who recorded his autobiographical reminiscences:

> For the first time during my management of the Lane the pantomime was a miserable failure. The cause was not far to seek. There was a certain troupe of pantomimists whom, for years, I had made the all-absorbing feature of my Christmas fare. For years I had kept them on, at continually increasing salaries, until at length they became paramount. Their ability was unquestionable; but the truth is, they were played out, and their attraction was over.[21]

Although *The Times* (27 December 1878) reported of *Cinderella*, 'the dialogue was sometimes less than sparkling and the puns with which it was littered, not above average', there was the customary praise from the critics for Beverley's scenery, the ballets and the Vokes family. The truth is that Chatterton's financial situation had become too precarious. He admitted to having made a loss of £7,000 on the season leading up to *Cinderella*. But according to Edward Stirling, the Drury Lane stage manager, he owed £36,000 when the crash came.[22] He admitted to having borrowed money at 'usurious interest'.[23] When Chatterton found himself unable to pay the salaries of the cast, the Vokes family went on strike and Chatterton had no alternative but to close the theatre on

4 February 1879. William Beverley wrote indignantly to *The Daily News* (15 February, 1879) implicitly denouncing the Vokeses for demanding their 'pound of flesh' and claiming that the other staff would have been willing to carry on accepting reduced salaries until Chatterton could recoup his finances. But the Vokes' action had resulted in the whole staff being put out of work, and Chatterton losing the lease of Drury Lane. Chatterton's friends rallied round and several benefits were held to raise money for him. The Gatti brothers, who took Covent Garden for the Christmas season, engaged him as their manager for the 1880–81 pantomime season. He produced the pantomime *Valentine and Orson* but a quarrel with the Gattis about his benefit at the end of the season led to him leaving Covent Garden. John Coleman took up the story: 'The year after he left Covent Garden, yielding to bad advice, Chatterton was unfortunately induced to tempt fortune again by embarking on management at Sadler's Wells Theatre. The result was most disastrous. The receipts at no period of the season ever approached half the expenses. Nor was this the worst. As there was no capital to work upon, the productions were of such a character as to be absolutely ruinous to his reputation.'[24] He opened on 29 September 1881 and the theatre closed abruptly on 4 February 1882. His last throw had been the pantomime, *Ali Baba and the Forty Thieves*, co-written by Chatterton himself. Chatterton, drawing on his bitter experience of a lifetime of lawsuits, made the forty thieves members of the legal profession. John Coleman who went to see it recorded: 'It was, I think, the saddest sight I have ever witnessed. An empty house, and a shabby, tawdry show which would have discredited a respectable barn … It was evident he had reached the bottom of the hill.'[25] Chatterton never returned to management, but sought to eke out a living as a reciter of passages from Dickens novels before dying of bronchitis in 1886 aged only 52.

Whether the Vokes family were quite as played out as Chatterton suggested is open to question. *Bluebeard*, in which they starred for Harris, elicited from *The Times* (27 December 1879) the verdict that the acting throughout was 'excellent', many of the songs and dances were encored and the leading performers were called several times to take a bow. It was staged on 'a scale of great magnificence', its chief merits being 'scenic display and burlesque humour'. In place of Beverley, Harris had employed the team of Harry Emden, William Telbin and F. Fenton. But Harris was determined to be rid of the Vokeses, telling journalist H. G. Hibbert that he dreaded their 'tyranny'[26] and although Blanchard supplied the script for the next show, *Mother Goose and the Enchanted Beauty* (1880–81), Beverley and the Vokeses were gone.

What is remarkable in the light of Chatterton's subsequent comments about them is that when he transferred to Covent Garden to produce the pantomime *Valentine and Orson or Harlequin and the Magic Shield* by Frank Burnand, he immediately engaged the Vokes family and Drury Lane veterans William Beverley as scene-painter and John Cormack as ballet-master. *The Times* (28 December 1880) praised the pantomime for its 'elaborate painting and splendour' and the performance of the Vokes family: 'Their dancing is as ever of the lightest, the wildest and yet most finished in character.' But the pantomime fell victim to the weather. Chatterton recalled: 'Our pantomime there made a great mark, and we should doubtless have done big business had it not been for the dreadful snow-storm which ruined half the managers in town and country. Doubtless you remember: half a dozen West End theatres were closed at a moment's notice in consequence of the streets being impassable.'[27]

The Vokes family were absent from West End pantomime for the next two years, returning in *Little Red Riding Hood* at Her Majesty's Theatre. It has something of the feel of a comeback both for them and for 'the traditional pantomime'. *The Times* (27 December 1883) rhapsodized that it had been produced with 'scrupulous regard for pantomime tradition in the treatment of fable' and 'marvellous fertility of resource in the matter of episode and spectacular display'. The critic argued that the artistic level attained by *Little Red Riding Hood* and *Cinderella* at Drury Lane would be 'satisfactory to those who have deplored the corruption of Boxing-day morals through the influence of the music hall. Here, and at Drury Lane, the youthful spectator is at once transported from a prosaic world into the realm of fancy and kept there while the librettist, the scene-painter, the stage-carpenter and the stage manager combine to awe and dazzle him with alternate doses of the supernatural, the grotesque, the mysterious and the beautiful'. *The Era* (29 December 1883) reported:

> Mr Fred Vokes … accomplished feats astonishing even his oldest admirers … Mr Fred Vokes has lost nothing of his striking ability as a dancer and pantomimist, while he has certainly gained new power as a burlesque actor … Miss Jessie Vokes … made a very amusing schoolmistress … As for Miss Victoria Vokes, she was the very life and soul, the central figure of the pantomime, the moving spirit of all. Never have we seen her so full of fun and energy.

But it was virtually their last hurrah. Rosina had left the troupe in 1877 when she married. Jessie, who had been 'a great favourite' of Blanchard, died on 4 August 1884 aged only thirty-three and Fred died on 3 June

1888 aged forty-two. Meanwhile Augustus Harris saw off his rivals. Covent Garden ceased producing Christmas pantomimes after 1882 and Her Majesty's did not return with another pantomime in 1884.

By the 1870s the nature of pantomime was changing, with the role of the Harlequinade increasingly curtailed, prose replacing the rhyming couplets and most significantly the interjection into the stories of music hall comedians and their acts. Augustus Harris Senior (1825–1873), who for twenty-seven years was associated with the stage and general management of Covent Garden, was retrospectively accused of beginning the 'vulgarization' of the pantomime and specifically introducing music hall performers into West End pantomime. The key date was said to be 1871 when music hall star G. H. Macdermott appeared in *Bluebeard* at Covent Garden. Soon music hall performers (among them James Fawn, Herbert Campbell and Harry Nicholls) were performing in pantomimes in both the West End and the East End.[28] It was Harris' son, Augustus Harris Junior, who set the seal on this development after he took over Drury Lane in 1879. The Christmas pantomime now became a combination of music hall knockabout (with the much-loved team of Dan Leno and Herbert Campbell packing in the audience at Drury Lane) and costly and lavish spectacle, in particular long processions on such themes as the Kings and Queens of England, nursery rhyme characters and famous beauties of history. So from the 1870s battle was joined for the soul of pantomime and the purity of fairyland.

It was after a hiatus of eight months that the lease of Drury Lane was awarded to the ambitious, enterprising Augustus Harris Junior (1852–96), then only twenty-seven. He began his regime at Drury Lane with a production of *Henry V* on 6 November, 1879. He at once applied to Blanchard for a pantomime script. But Blanchard was extremely wary of Harris, knowing his style from previous productions, and rather than write a new one, offered him an old script of *Bluebeard* which he had written with T. L. Greenwood under the pseudonym of The Brothers Grinn. It was extensively revised without credit by the writer and journalist Horace Lennard, who became Harris' literary adviser. Interestingly Blanchard did not see the pantomime until 12 February and recorded 'find the text set at nought'.[29]

In his regime at Drury Lane, Harris created a winning formula: with the annual autumn melodrama, often on an imperial theme, followed by a spectacular pantomime. He often collaborated on the melodrama and pantomime scripts and supervised every aspect of the production. He became the greatest impresario of his age, earning a knighthood before his early death at the age of forty-four.

Harris sought each year to outdo himself in the size and splendour of the pantomime. Cast size was increased. In *Forty Thieves* (1886–7) each of the thieves was given his own retinue. The processions and the ballets became longer and more populous. The number of music hall performers was stepped up. The spectacle was extended. This led to disaster in the 1882–3 pantomime *Sindbad*. As journalist J. B. Booth recalled:

> Nobody who saw it will ever forget the first night of ... *Sindbad the Sailor*. Several of the scenes had never been set when the curtain rose on that memorable performance, and some of the elaborate mechanical effects had never been tested. So, after the big bird which was to carry off Sinbad, refused to fly, and the great whale lay flabbily immovable on the stage, Harris had to come forward ... to ask the indulgence of the audience ... And then Charles Lauri, who played the part of a poodle, frisked about in the canvas waves, to draw attention from the carpenters who were dissecting the great whale, and as soon as there was room in front of it, a cloth was dropped, so that the comedians, Arthur Roberts, Jimmy Fawn, Harry Nicholls, and Herbert Campbell could 'gag' while the leviathan was being removed piecemeal.[30]

While the visuals and the ballet earned their customary critical praise, the general verdict was that Harris had tried to pack in too much and had lost sight of the story.

The criticisms of *Sindbad* seemed to have induced a temporary change of heart in Harris. For the 1883–4 pantomime *Cinderella* caused *The Times* (27 December 1883) to comment approvingly: 'the happy idea has been acted upon of eliminating to a great extent the sordid music-hall element which had mixed itself up with the amusements of Boxing Day and of keeping the fairy tale to the region of fancy to which it belongs'. William Beverley had been summoned back to join the scene-painting team of Henry Emden, William Telbin, W. Perkins and T. W. Grieve and there was praise for the fairy conclave in the moonlit glen ('a beautiful scenic effect') and the fairies' ballet ('most effective'). *The Era* (29 December 1883) concluded that 'At no previous period has so much care and attention been paid to the brilliancy and tastefulness of the decorative accessories. Costumes and scenery have now reached a degree of excellence of design and elaborateness of execution never anticipated even a quarter of a century ago.'

But Harris was back with size and spectacle in 1884–5 and not one but two grand processions in *Dick Whittington and his Cat. The Times* (27 December 1884) saw the keynote of the show as a blend of spectacle ('gorgeous scenery and fantastic pageants') and fun ('sprightly and humorous acting'). But the critic added: 'what may be called the dramatic interest of the story is necessarily to a great extent submerged in a vast and ever-changing panorama of civic and barbaric splendour'. Blanchard's health began to break down in the 1880s and he lacked the energy to resist the Harris revolution. Harris dubbed him 'the old man of the sea' as he clung on to the bitter end, but he was conscious of the publicity value the Blanchard name still carried. Blanchard's diaries record his increasing distress at the way in which his scripts were distorted and adulterated by Harris to accommodate the music-hall favourites. Whereas he regularly recorded of the pantomime productions in the pre-Harris years 'goes off well' or 'goes off very well', from the start of the Harris regime he was unhappy. On 19 December 1881 he wrote: 'Look over the ghastly proofs of my Drury annual (*Robinson Crusoe*), in which I find my smooth and pointed lines are turned into ragged prose and arrant nonsense. Consider the payment made to me as an equivalent for the harm done to my literary reputation, and shall henceforth look upon it in that light alone.'[31] On 26 December 1882 he wrote of *Sindbad* at Drury Lane: 'though exquisitely got up, it is a very dreary music-hall entertainment; and for the misprinting and grossly interpolated book I am in no way responsible. It is deservedly hissed at various portions – hardly anything there as I had intended, or spoken as I had written; the music hall element crushing out the rest, and the good old fairy tales never to be again illustrated as they should be.'[32] So it goes on until the final despairing entry on 21 December 1888: 'Go through the distasteful librettos of *Babes in the Wood* giving Augustus Harris permission to use my name in consideration that he uses some lines of my old annual of seventeen years ago'.[33] Just as his first pantomime for Harris had been a revival of an old script, so too was his last one. It seems that leading Drury Lane comedian Harry Nicholls had rewritten both *Dick Whittington* (1884) and *Aladdin* (1885) without credit. But *Babes in the Wood* bore the credit 'written by E. L. Blanchard, Augustus Harris and Harry Nicholls'. It was in many respects the end of an era. William Beverley died in the same year as Blanchard, on 17 May, 1889 at the age of seventy-eight. Symbolically Dan Leno made his Drury Lane debut in *Babes in the Wood* remaining a fixture in the pantomimes there until his death in 1904, creating the modern pantomime dame and being idolized by audiences.

One of the champions of traditional pantomime was John Ruskin who approvingly referred to the 'Arcadias of Pantomime', seeing them as representations of an ideal or idealized world.[34] He compared church and pantomime, 'these two theatrical entertainments where the imaginative congregations still retain some true notions of the value of human and beautiful things ... also they retain some just notion of the truth, in moral things'.[35] He saw evidently these values in Blanchard's pantomimes *Jack in the Box* and *The Dragon of Wantley*, which he saw respectively on 22 January 1874 and 18 January 1871. He said in *Fors Clavigera* 39 (March 1874) that *Jack in the Box* had 'everything I want people to have always, got for them, for a little while' – namely the combination of scenic beauty and moral instruction.[36]

Ali Baba and the Forty Thieves at Covent Garden, which he saw in 1867, had the opposite effect. He denounced it in a letter to the *Pall Mall Gazette* (1 March 1867) published under the heading 'The Corruption of Modern Pleasure'.[37] This pantomime, written by Gilbert A'Beckett, subordinated the fairy tale elements of the story to contemporary references. The thieves' cave was an up-to-date London club complete with billiard tables and the forty thieves were played by cigar-smoking, scantily clad girls. Ruskin was appalled, regarding it as a desecration of womanhood and a betrayal of the fairytale pantomime. Blanchard shared Ruskin's distaste, writing in his diary after a visit to Covent Garden that the story was 'coarsely treated – all legs and limelight'.[38] Unsurprisingly the show was produced by Augustus Harris Senior, the pioneer of pantomime 'vulgarization'.

Voice after voice was raised in defence of the old pantomimes. The critic W. Davenport Adams wrote an article entitled 'The Decline of Pantomime' in *The Theatre* (1 February 1882).[39] He denounced the vulgarization of the pantomime by the introduction of the music hall element, which led to 'impropriety of word, gesture and business which makes so much of our pantomime unsuited to the youthful ear and eye and ... unpleasant to all people of whatever age, who possess great taste and feeling'. The music hall songs, gags and dances were, he thought, all out of place and the spectacle of men dressed as women and scantily-dressed women posing as men was gratuitous and deeply distasteful to set before children. In his book *The Pantomimes and All About Them* (1881), rather incongruously dedicated to Augustus Harris, whom many held responsible for the deterioration of the much-loved form, Leopold Wagner began by complaining:

> There was a time, alas! when Pantomime was something very different to the now-a-day entertainment of that name, which has been described as 'a mass of insane absurdity and senseless incongruity',

> a time when reason and fanciful invention were called into requisition, to produce results as dignified as they were pleasing. Those were the days when visiting the theatre during the coldest nights of January might be regarded as a real treat; when the Pantomime was built upon a story or legend, intelligible to the merest child; when genuine delight was depicted on every countenance; when the jokes were new and practical; when acting was in earnest, and singing rendered with due regard to vocal harmony; when knockabout niggers, clog dancers, gymnasts, contortionists, Whitechapel songsters, and other music hall 'novelties' were not considered indispensable for success; and when pageants and processions, realistic representations of farm-yards with their live stock, cataracts of real water, and extravagant ballet scenes, with the lime-light directed upon an array of palpably naked thighs, which decency should require to be covered with skirts – were as yet unheard of.[40]

The journalist Charles Dickens Jnr. writing 'On the Decadence of Pantomime' in *The Theatre* (27 January 1896) complained: 'The simple fact is that the pantomime of to-day is not only not pantomime at all, but has no connection whatever with it.' He recalled with nostalgic regret the Vestris extravaganza with its 'beautiful' final scenes by William Beverley that led managers to compete in visual extravagance and 'the delightful and never-to-be forgotten Vokes family – the nimble and humorous Fred, the charming Victoria ... and the irresistible Rosina'. But then came the deluge. 'The floodgates of the music hall opened and all that was agreeable about "the grand comic Christmas pantomime" was promptly and effectually drowned out.' And then began the period of 'hopeless, inane, and offensive vulgarity all over the country' which, he admitted, had been 'highly remunerative to performers and managers alike'. He concluded sadly: 'The old English pantomime is dead.'[41]

Augustus Harris hit back at such criticism and robustly defended the deployment of spectacle in the theatre in an article entitled 'Spectacle' in *The Magazine of Art* (1889). He rejected the calls for the revival of the old Harlequinade, dismissing it as primitive and visually wretched:

> The dear old ladies and gentlemen who now insist that this class of performance should be played to day, because it met with the approval of their generation, forget that the School Board was not invented when they were young. Whereas in the olden times a

> pantomime ran a few weeks, it now runs so many months. In some of our provincial towns the harlequinade is now dispensed with, and in all others reduced to the smallest limits. As in London so in the provinces, the appearance of the clown is the signal for departure.[42]

So for Harris the key to theatrical change was the increased level of education of the audience. He argued that it was not sufficient just to spend money:

> A spectacular theatre must be, so to day, the trysting-place of all the arts. The work itself must be a labour of love, of perseverance, and of pluck; the co-operation of the most accomplished masters of the various arts should be secured ... It is, indeed, a question whether all the exquisite colourings and delightful combinations are fully appreciated, except by a small and highly-cultivated minority. Why, then, many ask, take the trouble to do the thing properly if not appreciated? Why not follow the old Boucicaultian managerial axiom, 'never try to educate your audience'? Because the minority is fast becoming the majority, thanks to the march of education ... the liberal expenditure of money in itself has little to do with an artistic result, but Art is a very extravagant and lavish mistress ... those who talk so much of the love of art whilst producing very little that is artistic, are generally those who are either unable or unwilling to make such financial sacrifices at the shrine of the goddess as she demands.

So Harris is defending his lavish pantomimes in terms of art and education, almost a Ruskinian position paradoxically for someone who traditionalists accused of vulgarizing and undermining the form.

Jimmy Glover, musical director at Drury Lane, writing in 1911, put it in perspective: 'It is ridiculous for people to state that pantomime has declined. It has not declined; it has, however, changed, and increased its public ... Whatever the criticisms may be, the intent of the present pantomime provider has always been good. We live in an age of splendour, luxury and comfort. Therefore the best in spectacular display, in talent, in music, and interpretation is provided.'[43] It is clear too that audiences loved them, and whatever the laments of the traditionalists, the box office had the final say. The Augustus Harris formula was followed all over the country and entered a new phase which combined spectacle, pantomime and music hall as the fairy world of Blanchard and Ruskin was decisively defeated.

Notes

1. Diane Purkiss, *Troublesome Things: a history of fairies and fairy stories* (London: Allen Lane, 2000), p. 220.
2. Jack Zipes, ed., *Victorian Fairy Tales* (London: Methuen, 1987), pp. xiii–xxix.
3. Ibid., p. xvii.
4. Charles Dickens, *Dickens' Journalism* vol. 3, ed. Michael Slater (London: J. M. Dent, 1998), pp. 167–8.
5. Richard Altick, *Paintings from Books* (Columbus, Ohio: Ohio State University Press, 1985), p. 263.
6. Iain Zaczek, *Fairy Art* (London: Starfire, 2001); Jane Martineau, ed., *Victorian Fairy Painting* (London: Merrell Holberton, 1997).
7. E. T. Cook and Alexander Wedderburn, eds., *The Complete Works of John Ruskin* (London: George Allen, 1903–1912), vol. 3, pp. 327–49.
8. Ibid., vol. 33, p. 327.
9. Tracy C. Davis, 'What are Fairies For?' in Tracy C. Davis and Peter Holland, eds., *The Performing Century* (Basingstoke: Palgrave Macmillan, 2007), pp. 32–3.
10. William Archer, *The Theatrical 'World' of 1893* (London: Walter Scott, 1894), p. xiii.
11. Clement Scott and Cecil Howard, eds., *The Life and Reminiscences of E. L. Blanchard* (London: Hutchinson, 1891), vol. 2, p. 599.
12. Ibid., p. 586.
13. Gerald Frow, *'Oh, yes it is' – a history of pantomime* (London: BBC Books, 1985), p. 124.
14. Scott and Howard, *The Life and Reminiscences of E. L. Blanchard*, vol. 1, pp. v–xv.
15. Ibid., pp. 148, 273.
16. J. R. Planché, *Recollections and Reflections* (London: Tinsley Brothers, 1872), vol. 2, p. 135.
17. Scott and Howard, *The Life and Reminiscences of E. L. Blanchard*, vol. 1, p. 318; Edward Wakeling, ed., *Lewis Carroll's Diaries* (Luton: the Lewis Carroll Society, 1999), vol. 5, p. 127.
18. H. G. Hibbert, A *Playgoer's Memories* (London: Grant Richards, 1920), pp. 101–2.
19. John Coleman, *Players and Playwrights I Have Known* (London: Chatto and Windus, 1888), vol. 2, p. 345.
20. Ibid., p. 358.
21. Ibid., p. 376.
22. Edward Stirling, *Old Drury Lane* (London: Chatto and Windus, 1881), vol. 1, p. 317.
23. Coleman, *Players and Playwrights*, vol. 2, pp. 345, 376.
24. Ibid., pp. 385–6.
25. Ibid., p. 386.
26. Hibbert, *A Playgoer's Memories*, p. 78.
27. Coleman, *Players and Playwrights*, vol. 2, p. 382.
28. A. E. Wilson, *Pantomime Pageant* (London: Stanley Paul, 1946), pp. 57–8.
29. Scott and Howard, *The Life and Reminiscences of E. L. Blanchard*, vol. 2, p. 498.

30. J. B. Booth, *The Days We Knew* (London: Werner Laurie, 1943), p.115.
31. Scott and Howard, *The Life and Reminiscences of E. L. Blanchard*, vol. 2, p. 528.
32. Ibid., pp. 541–2.
33. Ibid., p. 269.
34. Cook and Wedderburn, *Works of Ruskin*, vol. 27, p. 256.
35. Ibid., vol. 28, p. 52.
36. Ibid.
37. Ibid., vol. 17, pp. 333–8.
38. Scott and Howard, *The Life and Reminiscences of E. L. Blanchard*, vol. 1, p. 335.
39. W. Davenport Adams, 'The Decline of Pantomime', *The Theatre* n.s. vol. 5 (February 1882), pp. 85–90.
40. Leopold Wagner, *The Pantomimes and All About Them* (London: John Heywood, 1881), p. 9.
41. Charles Dickens Junior, 'On the Decline of Pantomime', *The Theatre* vol. 27 (January 1896), pp. 21–5.
42. Augustus Harris, 'Spectacle', *The Magazine of Art*, vol. 12 (1889), pp. 109–13.
43. James Glover, *Jimmy Glover: His Book* (London: Methuen, 1911), pp. 158, 161.

2

'Arcadias of Pantomime': Ruskin, Pantomime, and *The Illustrated London News*

Sharon Aronofsky Weltman

John Ruskin enthusiastically attended the theatre throughout his life, particularly enjoying pantomimes. As a child, Ruskin found every pantomime 'a matter of intense rapture';[1] as an old man, he still enjoyed it as one of his few untarnished delights.[2] But in addition to his own pleasure in theatrical entertainment, because pantomime reaches all classes, it is for Ruskin a perfect popular cultural reference to illustrate his points in urging a more just society. His books for working class readers *Fors Clavigera* and *Time and Tide* both depend on analyses of fanciful pantomimes to make their arguments. In these and other works, Ruskin wants most for theatre to provide a link to the world outside, not, as one might think, to represent that world more accurately or realistically and also not simply to function as a didactic tool telling moral stories, but rather to make manifest in performance the possibility of other ways to act. In other words, as I have written elsewhere, for Ruskin the enactment of a fictional existence, identity, or idea on stage momentarily realizes it.[3]

I expand on my original argument here, comparing Ruskin's reactions to Victorian pantomime with engravings from *The Illustrated London News*. Juxtaposing these periodical representations to Ruskin's rousing word paintings of the same thing exposes the significance and complexity of Ruskin's responses to pantomime and the uses he makes of it. The illustrations exhibit some of the same tensions that Ruskin's commentary reveals, such as the beauty of the fairy-tale universe inside the theatre versus the misery of the poor outside, or the labour of pantomime child actors versus their portrayal of an ideal. But these depictions do not address Ruskin's special concern with using pantomime to imagine and create a better world.

In *Fors Clavigera* Letter 39 (March 1874), Ruskin jokes that he cannot tell the difference between pantomime and reality.[4] He has just bought tickets for his fifth visit to see the current season's *Cinderella* at Hengler's Circus; he also buys tickets for the Drury Lane pantomime *Jack in the Box* for the following day.[5] Here he invokes scenes of both poverty and entertainment in London that he knows his multi-class readership will recognize in order to place before them a contrast between what ought to be and what is:

> [D]uring the last three weeks, the greater part of my available leisure has been spent between Cinderella and Jack in the Box; with this curious result upon my mind, that the intermediate scenes of Archer Street and Prince Street, Soho, have become to me merely as one part of the drama, or pantomime, which I happen to have seen last; ... I begin to ask myself, Which is the reality, and which the pantomime? Nay, it appears to me not of much moment which we choose to call Reality. Both are equally real; and the only question is whether the cheerful state of things which the spectators, especially the youngest and wisest, entirely applaud and approve at Hengler's and Drury Lane, must necessarily be interrupted always by the woeful interlude of the outside world.[6]

By making everyday life 'part of the drama', Ruskin turns pantomime into the overarching truth, while our supposedly real lives are downgraded to woeful interludes within the play. Far from failing in mimesis (as in *Modern Painters* he had accused opera of doing[7]), pantomime provides the ideal that the real should imitate. Ruskin describes the actors in *Cinderella* 'all doing the most splendid feats of strength, and patience, and skill ... [T]he pretty children [are] beautifully dressed, taught thoroughly how to behave, and how to dance, and how to sit still, and giving everybody delight that looks at them.'[8] In contrast, Ruskin complains that 'the instant I come outside the door, I find all the children about the streets ill-dressed, and ill-taught, and ill-behaved, and nobody cares to look at them.'[9]

He makes the contrast even more fiercely in *Fors Clavigera* Letter 14 (February 1872), presenting a much more gruesome demarcation between pantomime and street children. He acknowledges that the loveliness of 'the pastoral scene in a pantomime' must 'remain somewhat incredible' to his readers,[10] but nonetheless he wants pantomime beauty to become believable and real enough to replace:

> [t]hose children rolling on the heaps of black and slimy ground, mixed with brickbats and broken plates and bottles, in the midst of

> Preston or Wigan, as edified travellers behold them when the station is blocked, and the train stops anywhere outside, – the children themselves, black, and in rags evermore, and the only water near them either boiling, or gathered in unctuous pools, covered with rancid clots of scum, in the lowest holes of the earth-heaps.[11]

Ruskin rallies his readers to recognize injustice by demonstrating the tragic contrast between child actors (who represent the acme of strength, skill, beauty, and exuberant youth) and what amounts to industrial crime against children. Yet he never mentions that the pantomime players' exemplary behaviour is an act, their beautiful clothes costumes, and the delight they provide simply a way to earn money.[12]

Ruskin's horrified concern for street children and his unmitigated admiration for the children on stage suggests a tension that is illuminated by an image from a few years earlier, December 1867: 'Engaging Children for the Christmas Pantomime at Drury Lane Theatre,' by Mason Jackson, art editor of *The Illustrated London News* from 1860 to 1890. (Figure 2.1) In this image, a multitude of youngsters vying for parts in a pantomime (probably Blanchard's *Faw-Fee-Fo-Fum; or, Jack the Giant Killer*) crowd around the stage door of the Theatre Royal at

Figure 2.1 Mason Jackson, 'Engaging Children for the Christmas Pantomime at Drury Land Theatre', *Illustrated London News* 51 (7 December 1867): 612

Drury Lane, where Samuel Phelps currently stars in *The Doge of Venice*. The principal impression created by the scene is of a superabundance of activity in a crowded public space. The architectural perspective creates a sense of deep focus, carrying our vision all the way back to the waiting cab at the end of the colonnade; the sheltered sidewalk is depicted as much wider than the space really is, reinforcing this sense of multitude. Victorian pantomimes could have close to five-hundred supernumeraries; such a huge number is figured here by at least ninety children actually pictured within the frame, spilling outward beyond its view.[13]

Many tiny dramas play out across the illustration. They emphasize the aspirants' identity as children and as working-class theatre professionals. An adolescent girl in the left foreground watches the scene with her back to us, her legs well turned out like a ballerina in sturdy second position. She carries a baby on one shoulder and a basket on her other arm. Some of the children in her vicinity look anxious as they form a circle around boys starting to fight, but others seem excited, perhaps more about the fight than the audition. Many are practising dances or acrobatics. Two girls twirl together on the right, feet pointed and hair flying; a boy in the foreground stands on his hands in precise imitation of the poster beside the stage door; a friendly dog and several children gaze eagerly at him in the illustration's brightest light, making him the focal point. A teenager on the right sits in exhaustion while clutching a broom; a weary-looking girl rests her hands on the sweep's shoulders. In striking contrast to this pair, to the farthest right, a well-dressed boy and his young mother, whose restraining hand rests on the boy's shoulder, watch the scene, standing in for the periodical's middle-class readership. The tired girl and the rich boy look directly at each other, emphasizing their differences in class and occupation but connecting them as children. A boy in the foreground to the left bends under the weight of the shoe-shine box he carries on his back, suggesting that he has more than one job. Both a stage and a street child, he conflates Ruskin's two categories. He claps his hands to get a small short-haired dog's attention, but this dog, cringing with his tail between his legs, seems less playful, the closest the illustration comes to suggesting negative criticism of the scene.

The major areas of focus form miniature spectacles that represent the ways in which these young performers will entertain their audiences once the pantomime is mounted inside the theatre, often with the same or similar working-class spectators enjoying the show. The crowded image stresses the bustle, energy, and liveliness of these children, but also very much their professionalism. They clearly hope to get work.

They carry their tools of the trade and, evidently from the basket held by the baby-toting girl on the left, their lunch. The brief accompanying article in this issue of *The Illustrated London News* reports more unfavourably than the visual representation (or Ruskin), commenting that 'hundreds of poor families' 'yearly depend on this incidental gain ... of a few shillings' when they 'let their children be hired' as imps or fairies in Christmas pantomimes.[14] Yet the liveliness and general charm of the full-page image belies the tiny article's critique, implying that it is after all a good thing that so many poor families 'let their children be hired' in service of public entertainment.

The children pictured by Jackson differ markedly from those in R. Taylor's full-page 'Going to the Morning Performance of the Pantomime', *The Illustrated London News* (January 13, 1872) (Figure 2.2). These youngsters are on their way to see *Tom Thumb* at the Theatre Royal in Drury Lane, the same theatre shown a few seasons earlier in the previous illustration. A carriage conveys a large family with two young children, a fair-haired boy up front on his father or older brother's lap and a little blonde girl inside, looking out directly at the viewer through her opera glasses; she will not see the horrors confronting rail passengers in Preston or Wigan. A mother and older children, and perhaps other

Figure 2.2 R. Taylor, 'Going to the Morning Performance of the Pantomime', *Illustrated London News* 60 (13 January 1872): 48

relatives, comprise the group. The jewels and fur-trimmed cloak, the quality of the father's dress coat, the shine on the coach and its lantern, the fact that even the men can go to the pantomime instead of to work during the daytime all suggest that these are well-to-do theatre-goers.

Linking the two pictures is the children's excitement, obvious in the boy's leaning back eagerly to ask his father a question and the little girl's curiosity in observing the world around her through the novelty of lenses about to be used in the theatre. While the working-class children in the previous image hope to be employed, these children expect to be entertained. Their passivity is represented by their sitting and being carried instead of dancing or even standing, by the little girl's being entranced by receiving information through opera glasses, and the little boy's eagerness to be told something by his father. Unlike the active workers outside the theatre door in the previous image, all the children here are enclosed either by the protective walls of the carriage or by a parent's encircling arms and a warm blanket. This feel-good illustration suggests no critique of child labour.

Ironically, the article accompanying this illustration praises the institution of the matinée because it promotes the health of children in the audience by allowing them to go to bed on time:

> No boy or girl under ten years of age should ever be out of bed after 10:00 o'clock at night; and it is cruelty, though often meant for kindness, to offer the young people a momentary pleasurable excitement at the cost of their future soundness and vigour, both in mind and body … We are greatly pleased to know that the children going to the pantomime, who appear in our Engraving bent on enjoying themselves to the utmost, will be carried back to their home before six o'clock.[15]

Yet, like Ruskin, *The Illustrated London News* omits the point that the child performers in the pantomime must stay up late every night they perform in the show. As Tracy Davis points out, not until the Prevention of Cruelty to Children Bill of 1889 (amended in 1894 to include boys), were theatres prohibited from employing children under age 16 to work after 10:00 pm.[16]

Taken together, these two images and the accompanying text provide a revealing context for Ruskin's idealized child actors and his blindness to their status as labourers. Yet, while Ruskin ignores the material lives of the actors as working men, women, and children in the passage I quoted earlier from *Fors*, he is not always oblivious: when he was a young man, his mother, distressed that he had blown a hundred pounds 'on grapes, partridges, and the opera', sent him five pounds 'to make peace with Heaven' by donating to churches. He defied her by giving the sum instead to an 'overworked ballerina in Turin' because she 'did her work well always; and looked nice, – near the footlights'.[17] Much later, in the early 1880s, the aging Ruskin learned substantially more about the material reality of performing children through a friendship with the young Webling sisters, who gave widely admired public poetry recitations.[18] Perhaps this late acquaintance gave him a deeper understanding of child actor as child labourer, but in the 1860s and 70s, his experience of the stage remained one that highlighted differences between theatrical magic and the appalling reality outside the theatre.

Back to Letter 39 of *Fors Clavigera*: in an attempt to rally his readers to social action, Ruskin has just contrasted the beautiful world of stage children in *Cinderella* and *Jack in the Box* with the wretchedness suffered by street children. Related to that contrast, he goes on with dry humour to decry the audience's reactions to pantomime special effects. At Drury Lane's *Jack in the Box*, people enjoy both

> Mushroom Common, with its lovely mushrooms, white and grey, so finely set off by the incognita fairy's scarlet cloak; ... [and] Buttercup Green, with its flock of mechanical sheep, which the whole audience claps because they are of pasteboard, as they do the sheep in Little Red Riding Hood because they are alive; but in either case, must have them on the stage in order to be pleased with them, and never clap when they see the creatures in a field outside.[19]

The link between stage sheep and field sheep parallels Ruskin's contrast between stage children and street children. He collapses the differences between stage and street or field to inspire his readers to strive for social change. Yet Ruskin's frustration stresses the audience's reluctance to see theatrical performance as a connection to the outer world.

Before continuing with the rest of Ruskin's markedly ambivalent reaction to *Jack in the Box*, I would like to compare it to its verbal and visual depiction in the *Illustrated London News*, which again provides revealing context for Ruskin's remarks:

> The pantomime at the great national theatre is ... entitled 'Jack in the Box; or, Harlequin Little Tom Tucker and the Three Men of Gotham who went to Sea in a Bowl.' ... [M]uch ingenuity is displayed touching a fancy fairy fair held on Midsummer Eve every century. Imagine

> Felix, the Prince of the Fortunate Isles, awaiting the hour of midnight on Gotham-common, when, to his surprise, a number of mushrooms transform themselves into so many fancy stalls presided over by the fairies! One of these presents him with the talisman he wanted – namely, Jack in the Box – by means of which he has to cure of her idiotic conceit Princess Poppet, the beautiful but silly daughter of Cockalorum the Great, King of Cockaigne.[20]

The reviewer explains that Princess Poppet, played by Violet Cameron, is transformed for a time by the magic Jack into little Bo Peep as part of her cure; Prince Felix (also performed by an actress, as is typical in pantomime) joins his beloved disguised as Tom Tucker.[21]

In a follow-up review on 10 January, 1874, *The Illustrated London News* commends the excellence of these individual performers (whom we shall see that Ruskin also singles out for praise) particularly recognizing

> the merits of Mr. F. Evans, who sustains the part of Jack-in-the Box ... His spring into the air from the trap bespeaks the skillful acrobat, and his demeanor afterwards abounds in sportive suggestion ... Men and women become as excited as boys and girls.[22]

It is tempting, when reading these reviews or indeed E. L. Blanchard's script, to ask whether Ruskin's mixed emotions regarding *Jack in the Box* are connected to its abundant phallic imagery. There is plenty to work with: Cockalorum, King of Cockaigne, rules.[23] His daughter ultimately marries a fellow who controls the talisman of love, Jack in the Box, which leaps up to amazing heights, whose primary purpose is to go into and out of a beautiful box. Even the much touted mushrooms have a reputation for increasing in size with surprising speed. Interpreting the pantomime as an extravaganza of phallic obsession might explain Ruskin's repudiation of it in Letter 60 of *Fors* (December 1875), almost two years later; there he writes that 'Three Wise Men of Gotham' (part of the subtitle of this play) was 'the very worst pantomime I ever saw, having from the mere intolerableness of its stupidity, so fastened itself in my memory that I can't now get rid of the ring in my ears'.[24] Certainly some of the rhyming dialogue may explain this criticism: for example, just before the scene in which the mushrooms arise and turn into the Fairies' Fancy Fair and Flower Show, the lead fairy Elfina says:

> You see at present everywhere among us,
> A mushroom, popularly called a fungus.

I do but strike my crutch, and with this plunge, I
Reveal the funny figures of the Fungi.[25]

But in *Fors* Letter 39, after having just seen the play, Ruskin's approbation seems as strong as that of *The Illustrated London News*. The point of his praise is to contrast the beauty of the performance with the misery of the industrial revolution. The audience at *Jack in the Box*

> can't have enough, any more than I can, of the loving duet between Tom Tucker and little Bo Peep: they would make the dark fairy dance all night long in her amber light if they could; and yet contentedly return to what they call a necessary state of things outside, where their corn is reaped by machinery, and the only duets are between steam whistles. Why haven't they a steam whistle to whistle to them on the stage, instead of Miss Violet Cameron? Why haven't they a steam Jack in the Box to jump for them, instead of Mr. Evans? or a steam doll to dance for them, instead of Miss Kate Vaughan? They still seem to have human ears and eyes, in the Theatre; to know *there*, for an hour or two, that golden light, and song, and human skill and grace, are better than smoke-blackness, and shrieks of iron and fire, and monstrous powers of constrained elements.[26]

His purpose is to stress the importance of the staged world not only as one depicting fairy-tale happiness, but also as a venue for the display of the performers' art and skill as compared to the despair, poverty, and ill-will beyond the theatre walls. Ruskin repeats both Carlyle's repudiation of Victorian machinery in 'Signs of the Times' and Dickens's promotion of Sleary's Circus over Bounderby's Coketown in *Hard Times*, but in contrast to Dickens, Ruskin sees in the theatre not only an opportunity to cultivate fancy but also a manifestation of a realizable Ideal and a path to genuine humanity.

The illustration of *Jack in the Box* shown here is from *The Illustrated London News* 64 (10 January, 1874): 28, accompanying the magazine's follow-up review quoted above (Figure 2.3). The picture's main focus is on the high acrobatic leap of Jack, whose portrayal by Fred Evans earned such praise from *The Illustrated London News*, in addition to Ruskin's admiration in *Fors*. Here Jack springs to a seemingly impossible height out of the box. The artist D. H. Friston draws our attention to Jack by the backward slant of the two main foreground figures on either side leaning backward: on the left, King Cockalorum, recoiling in amazement; on the right, Prince Felix demonstrating the magical qualities of his gift.

Figure 2.3 D. H. Friston, 'Scene from 'Jack in the Box' at the Drury Lane Theatre', *The Illustrated London News* 64 (10 January 1874): 28

Princess Poppet examines her present with marked interest from the far left. The crowd surrounding the central action is diverse in appearance, including soldiers, courtiers, children, animals, and big-headed political caricatures of Disraeli and Gladstone. According to the published script, the cast includes a 'host of juvenile auxiliaries'. [27] Yet, while not exactly an Arcadia (no hint even of Bo Peep's sheep), this illustration offers no social critique, only an entertaining display of talent.

While Ruskin sees the pantomime as manifesting the True Ideal,[28] a reality more real than the world outside the theatre and a model for the so-called real world to emulate, a grim reversal remains a possible threat. In *Fors* 14, where we saw Ruskin contrast the pastoral loveliness of the Christmas pantomime with the ragged children of Preston 'covered with rancid clots of scum', he goes on furiously to ask his readers:

> why do you not paint these [children] for pastime? Are they not what your machine gods have produced for you? … Why do you not rejoice in these; appoint a new Christmas for these, in memory

of the Nativity of Boilers, and put their realms of black bliss into new Arcadias of Pantomime – the harlequin, mask all over?[29]

The bitterness of Ruskin's indictment here together with his emphasizing the seasonal link between the Christmas pantomime and Christianity suggests that Ruskin experiences the pantomime in something like spiritual terms, that the fairy-tale reality he proposes is not that different for him from religious truth. When discussing the production of *Cinderella* at Hengler's Circus in *Fors* 39, Ruskin equates the theatre and the church, where he points out that 'in these two theatrical entertainments – the Church and the Circus – the imaginative congregations still retain some true notions of the value of human and beautiful things', having neither 'steam-preachers nor steam-dancers'.[30] In trying to rouse anger against the horror of a mechanized society that neglects and abuses its children, Ruskin's rhetorical point depends upon knowing that his readers would never replace the Arcadias of Pantomime with such ugliness on stage, would never choose steam-dancers over live ones, and so should be convinced to reject industrial ugliness and exploitation in Preston, Wigan, and the streets of London as well. But lurking within this rhetorical question is an alternative answer: just as the church has become merely theatrical for Ruskin, so can the theatre lose its beneficial magic, and the pastoral retreat inside Hengler's or Drury Lane can become a glorification of the worst his culture offers instead of an alternative reality imagining and instantiating what could be its best.

Notes

1. *The Works of John Ruskin*. ed. E. T. Cook and Alexander Wedderburn (London: George Allen. New York: Longmans Green and Co., 1903–12), 35: 175.
2. Ibid., vol. 34, p. 669; vol. 37, p. 478.
3. See Sharon Weltman, *Performing the Victorian: John Ruskin and Identity in Theater, Science, and Education* (Columbus: Ohio State University Press, 2007).
4. See Van Akin Burd, *Christmas Story: John Ruskin's Venetian Letters of 1876–1877* (University of Delaware Press, 1991), p. 88, for a brief discussion of this moment.
5. Ruskin buys tickets to see *Cinderella* on 22 January 1874 at Hengler's Circus and *Jack in the Box* at Drury Lane on 23 January 1874. See Dinah Birch, ed. *Fors Clavigera* (Edinburgh: Edinburgh University Press, 2000), p. 348, n. 6. See Rachel Dickinson 'Ruskinian Moral Authority and Theatre's Ideal Woman', *Ruskin, the Theatre, and Victorian Visual Culture*, ed. Anselm Heinrich, Kate

Newey, and Jeffrey Richards (Basingstoke: Palgrave Macmillan, 2009), pp. 58–73, for more on Ruskin and Hengler's *Cinderella*.

6. Cook and Wedderburn, *The Works of John Ruskin*, vol. 28, pp. 50–1.
7. Ibid., vol. 6, pp. 390–3.
8. Ibid., vol. 28, p. 51.
9. Ibid., pp. 51–2.
10. Ibid., vol. 27, p. 255.
11. Ibid., p. 256.
12. For more about child actors as labourers, including a discussion of Jackson's engraving, see Tracy Davis 'The Employment of Children in the Victorian Theatre,' *New Theatre Quarterly* 2.6 (1986), pp. 116–35. Jim Davis also addresses this issue in 'Freaks, Prodigies, and Marvelous Mimicry: Child Actors of Shakespeare on the Nineteenth-Century Stage,' *Shakespeare* 2.2 (2006), pp. 179–93. Tracy Davis supplies information on Victorian legal responses to the labour of pantomime children: the Education Act of 1876, the Children's Dangerous Performances Act of 1879, and most significantly the Prevention of Cruelty to Children Bill of 1889.
13. For example, Michael Booth estimates that nearly five-hundred played in the 1886 production of *Ali Baba and the Forty Thieves*, in *Theatre in the Victorian Age* (Cambridge: Cambridge University Press, 1991), p. 35. Ruskin may not be exaggerating too outrageously when he mentions 'about four hundred and forty fairies" in addition to the forty thieves and their forty companions in an earlier production of *Ali Baba* at Covent Garden in 1867 (Cook and Wedderburn, *The Works of John Ruskin*, vol. 17, pp. 336–7; Weltman, pp. 30–6). See David Mayer, 'Supernumeraries: Decorating the Late-Victorian Stage with Lots (and Lots and Lots) of Live Bodies', in Heinrich, Newey and Richards, *Ruskin, the Theatre, and Victorian Visual Culture* (Basingstoke: Palgrave Macmillan, 2009), pp. 154–68, for more on Ruskin and supernumeraries.
14. *The Illustrated London News*, 7 December 1867, vol. 51, p. 608.
15. *The Illustrated London News*, 13 January 1872, vol. 60, p. 47.
16. See Davis, 'The Employment of Children in the Victorian Theatre,' *New Theatre Quarterly* 2.6 (1986), pp. 116–35.
17. Cook and Wedderburn, *The Works of John Ruskin*, vol. 35, p. 498.
18. Ibid., vol. 34, pp. 545–6; Joan Evans and John Howard Whitehouse, eds., *The Diaries of John Ruskin*, 3 vols. (Oxford: Clarendon Press, 1959), Vol. 3, p. 999; Tim Hilton, *John Ruskin: The Later Years* (New Haven and London: Yale University Press, 2000), p. 428.
19. Cook and Wedderburn, *The Works of John Ruskin*, vol. 28, p. 52.
20. *The Illustrated London News*, 27 December 1873, vol. 63, p. 623.
21. Like most pantomimes of the period, *Jack in the Box; or, Harlequin Little Tom Tucker and the Three Wise Men of Gotham*, pulls together a variety of stories. Folk tales about fools of Gotham are old, perhaps at the latest from the Renaissance. Washington Irving's having called New York 'the city of Gotham' in *Salamagundi* (1807) was in reference to these tales, and was a dig; the wise men of Gotham also appear in Thomas Love Peacock's poem 'Three Men of Gotham' from *Nightmare Abbey* (1818). Two of these traditional tales of Gotham form a large portion of the plot of this pantomime. Little Tom Tucker who sung for his supper comes from a nursery rhyme, later illustrated

by Ruskin's protégée Kate Greenaway. The Jack-in-the Box is far kinder than the character from Hans Christian Anderson's *The Steadfast Tin Soldier*.

22. *The Illustrated London News*, 10 January 1874, vol. 64, p. 27.
23. That Cockaigne also refers both to London and to a mythical land of plenty and that Cockalorum means *braggadocio* does not preclude their phallic significance.
24. Cook and Wedderburn, *The Works of John Ruskin*, vol. 28, p. 462. It's possible that Ruskin refers here to a different pantomime, but I have not been able to locate evidence of his having attended any other pantomime of this title. Birch also tentatively identifies *Jack in the Box* as the play Ruskin refers to here (*Fors Clavigera*, p. 359, n. 3).
25. E. L. Blanchard, *Jack in the Box; or, Harlequin Little Tom Tucker and the Three Wise Men of Gotham* (London: Tuck and Co., 1873), p. 18.
26. Cook and Wedderburn, *The Works of John Ruskin*, vol. 28, p. 52.
27. E. L. Blanchard, *Jack in the Box; or, Harlequin Little Tom Tucker and the Three Wise Men of Gotham*, p. 5.
28. Cook and Wedderburn, *The Works of John Ruskin*, vol. 6, p. 390.
29. Ibid., vol. 27, p. 256.
30. Ibid., vol. 28, p. 52.

3
Lewis Carroll, E. L. Blanchard and Frank W. Green

Richard Foulkes

Best known as the author of the *Alice* books, Lewis Carroll was an ardent theatre-goer throughout his adult life.[1] His extraordinary gifts – story-teller, mathematician, cleric, don and photographer – notwithstanding, Carroll's theatrical tastes were fairly typical of his social class at the time and he readily embraced that quintessentially Victorian entertainment the pantomime, though with some clear boundaries, principally those of respectability and decency, which he would not overstep. This led him to virtually confine his attendance at pantomimes to those by E. L. Blanchard and Frank W. Green, whom he trusted to avoid coarseness and at least resist and moderate the intrusion of music hall elements. In Blanchard's case his long association with Drury Lane entered a different phase under the management of Augustus Harris in the 1880s and Carroll ceased to attend. By then Blanchard's association with the Adelphi Theatre had ended and Frank W. Green was dead at the age of forty-eight.[2] For these reasons Carroll's pantomime-going was concentrated in a period of around a dozen years between 1874 and 1885, but brief though it was this might well be regarded as the golden age of pantomime for which happily Lewis Carroll can act as our guide.

E. L. Blanchard

A. E. Wilson dubbed E. L. Blanchard the 'Hero of a Hundred Pantomimes ... the most remarkable and sympathetic figure amongst all who contributed to Christmas entertainment'.[3] The mainstay of Blanchard's seasonal output was his Drury Lane 'Annual' (the term he preferred for pantomime) which occupied the stage of that theatre over Christmas and the New Year from 1852 to 1888.[4] Lewis Carroll attended five of

Blanchard's Drury Lane 'Annuals' beginning with *Little King Pippin* in December 1865 and concluding with *Cinderella*, which he saw twice in January 1878. In fact his first experience of the genre was Blanchard and Greenwood's (The Brothers Grinn) *Cinderella* at Covent Garden on 25 January 1865, which he described as 'a capital Pantomime'[5] so Blanchard may reasonably be credited with giving Carroll a taste for that form of entertainment. As Jeffrey Richards points out in his chapter on Blanchard at Drury Lane he was a tremendously prolific author (and journalist) who might have as many as five pantomimes at different theatres in the same year.

The Adelphi Theatre

Of these theatres it was the Adelphi in the Strand under the management of F. B. Chatterton that attracted Carroll in the mid-1870s. Blanchard's own interest in and knowledge of the Adelphi was evident in the lengthy article 'History of the Adelphi Theatre' that he contributed to the *Era Almanack* of 1877, in which he described the theatre's location as 'a conspicuous situation in one of the chief thoroughfares of central London, no theatre has ever been more eligibly placed for attracting the attention of provincial visitors'.[6] Although he divided most of his time between Oxford (Christ Church) and Guildford (where his sisters lived) Carroll was so often in London that he would hardly have considered himself a 'provincial visitor' and would have realised that though written under the guise of 'An Old Boy' the Adelphi pantomimes (they were never called 'Annuals') were Blanchard's work:

> The 'old boy' stands between the children who act and the children who are amused. He feeds the one and instructs the other. At his suggestion came these Pantomimes made for children, acted by children, and presented by children. He is the founder of this microscopic art which makes us believe more than ever in fairyland, and he is the faithful and constant friend of the good mothers who bless morning performances and are able to get the children safe in bed before the bell rings for supper.[7]

This extract highlights the binary nature of Victorian pantomime with on the one side the performing children, often earning vital wages for their families, and on the other those from more affluent circumstances for whom the occasion was a treat and a diversion. Child performers were not in themselves a novelty, for Blanchard's Drury Lane 'Annuals' were showcases for such youthful talents as Percy Roselle, Sallie Sinclair,

Figure 3.1 Scene from the children's pantomime version of *Little Goody Two Shoes* at the Adelphi Theatre, *Illustrated London News* (20 January 1877), 63

a hundred or more coryphées and the (maturing) Vokes family, but the difference was that at the Adelphi the cast consisted entirely of children. Clearly there was a huge market for this form of entertainment. Carroll had found the Adelphi's conventional offering *The Children in the Wood* in the 1874–75 season 'a poor Pantomime',[8] though Mr Fawn 'acquitted himself excellently, his imitation of Irving being much applauded',[9] but his enthusiasm for *Little Goody Two Shoes* (Figure 3.1) is abundantly evident from his 1877 diary entry:

> It was a really charming performance. The two Grattan children acted 'Goody' [Emilie] and 'Little Blue Boy' [Harry]: they seemed clever, but hardly worth the excessive praise they have received. Little Bertie Coote (about 10) was Clown, a wonderfully clever little fellow: and Carrie [Coote] (about 8) was Columbine, a very pretty graceful little thing – in a few years time she will be just *the* child to act 'Alice,' if it is ever dramatised. The Harlequin was a little girl named Gilchrist, one of the most beautiful children in face and figure, that I have ever seen.[10]

Carroll befriended the Cootes and ventured to the Surrey Theatre to see another member of the family, Lizzie, in a matinée performance of the pantomime which he considered 'very good as a whole, though rather spoiled by coarseness'. But, though matinée performances ensured that the younger members of the audience could be 'safe in bed before the bell rings for supper', performers of a similar age faced a long evening ahead of them before the curtain fell for the final time and they could wend their way home. The toll on Lizzie Coote, then aged fifteen, was apparent: 'She does not look as if the Pantomime season agreed with her, poor child, and she has the blue marks under her eyes which tell of late sitting-up.'[11]

As a result of the success of *Little Goody Two Shoes* there was an unprecedented extension of 'the Pantomime season' at the Adelphi. As the *Era* reported, 'the last "Children's Pantomime" – not so very many weeks since withdrawn [in April] from the Adelphi bill ... the prerogatives of the Christmas season are being overthrown'[12] with the staging of *Little Red Riding Hood* in August 1877, presumably as a school holiday treat. It was still running successfully in October when Carroll attended performances on two consecutive days, on 2 October in the evening and on the following afternoon. Whereas at Christmas the whole bill was customarily given over to the pantomime, with this unseasonal offering the management had evidently decided to hedge its bets and offer a double bill, initially with Paul Meritt's melodrama *The Golden Plough* and, by the time Carroll attended, with Boucicault's *After Dark*, an apt title given the *Era's* report:

> The 'Children's Pantomime,' it must be remembered, is played by children, and although a quarter to ten o'clock – the hour at which it commenced last Saturday – may be considered somewhat too late for the players and the spectators chiefly concerned, that cause for complaint will vanish with the morning performances which we are promised, and which are sure to be welcomed by those little people who are left in London and denied the pleasures of the seashore.[13]

Even the *Era*'s concern was primarily with the children in the audience, for whom matinées offered a more convenient option than evenings, rather than with those on stage who were of course required for both. Carroll's remark that 'The Pantomime was a good deal longer [at the matinée] than at night'[14] suggests that by then the evening version might have been compressed to facilitate an earlier finishing time for the benefit of the young performers. As Tracy Davis has pointed out the introduction of universal education and concern about child employment generally had raised awareness about conditions in the theatre.[15] Carroll's attention tended to be focused on particular families such as the Cootes and he recorded that Lizzie had now joined her siblings at the Adelphi albeit 'as "Area Jack," not a nice part for her, though she did it very cleverly'.[16] The attraction between Carroll and the Cootes appears to have been mutual if Bert's recollections are authentic: 'he was one of us, and never a grown-up pretending to be a child ... He was a born story-teller, and if he had not been affected with a slight stutter in the presence of grown-ups would have made a wonderful actor, his sense of theatre was extraordinary.'[17]

'Theatrical children always have a special attraction for me',[18] Carroll wrote to Tom Taylor, soliciting the dramatist's aid in support of Lizzie Coote, and he took the opportunity of the Adelphi pantomimes to befriend other young performers, especially Connie Gilchrist who had so entranced him in *Little Goody Two Shoes*. On 10 April 1877 he 'called on Mrs. Gilchrist and spent about half an hour with her and Connie'. He professed himself 'decidedly pleased with Connie, who has a refined and modest manner, with just a touch of shyness, and who is about the most gloriously beautiful child (both face and figure) that I ever saw. One would like to do 100 photographs of her.'[19] Carroll was certainly not alone in being spellbound by Connie Gilchrist, whether or not he succeeded in photographing her.[20] She was painted by Whistler performing her 'skipping-rope dance' and in 1892 married the 7th Earl of Orkney at All Souls, Langham Place with the Duke of Beaufort giving her away.[21]

Connie Gilchrist appeared in the next Adelphi Children's Pantomime *Robin Hood and his Merry Little Men* (1877–78) as did Bert and Carrie Coote. Carroll considered 'the children ... if possible more charming than ever',[22] attending six times and taking a succession of young companions. In this production he was particularly smitten by Sarah (Sally/Sallie) Sinclair as Cupid, whom he had previously admired as the Infant Cerito in *Little Red Riding Hood*. The Sinclairs were a theatrical family, the father (a comic actor) having changed his name from Scrivener.[23] Carroll corresponded with Sallie and her mother and, having effected an introduction through the Cootes, visited the family at 142 South Lambeth Road on 16 January 1878, pronouncing 'Sally' 'quite as beautiful by daylight as gaslight'.[24] Fate alas had misfortune in store for the Sinclairs, for Mrs Sinclair died in spring 1879 and Carroll dutifully penned letters of condolence to her husband and a family friend, Mrs J. Neate, to whom he expressed his concern for Sally ('the Stage has risks of its own, and most for those who have, like dear Sally, the dangerous gift of beauty', adding 'P. S. If any *money* help be needed, please let me know.')[25]

Carroll's engagement with the offstage lives of several of the leading children in the Adelphi pantomimes underlines the attractions and the precariousness of entering the profession at that age. The Cootes and the Sinclairs were both acting families for whose offspring a career in the theatre was probably the best prospect, since their earning capacity as children was a great bounty to their families. Connie Gilchrist came from a better-off background, her father being a successful 'civil engineer in good practice'.[26] However, but for her stage career, she would probably not have made such an advantageous marriage. Carrie Coote also 'married well' to barrister and MP Sir William George Pearce.[27] On the other hand the hazards of the profession manifested themselves in Lizzie Coote's poor health and Sallie Sinclair's enforced independence after her mother's death. Whatever the circumstances Carroll was steadfast, enlisting support for Lizzie Coote from Tom Taylor and Marion Terry, and offering cash as well as condolences to the Sinclairs.

In her study of children on the Victorian stage Anne Varty gives a full account of Blanchard's *Little Goody Two Shoes* at the Adelphi, making some comparisons with the script that he had produced for Drury Lane in 1862–63. She observes that the Harlequinade, which 'bears no links with the preceding drama', was 'similar to that of 1862' with its setting on 'The Deck of a Man o' War'.[28] In the intervening years, in a process that was to continue until its demise, the Harlequinade had diminished in importance and prominence. Blanchard who regarded himself as a bastion of tradition, especially against the incursion of music hall, endeavoured to maintain its position, aided by the careful supervision of Chatterton and John Cormack (of dances). As Carroll (no lover of the Harlequinade, in which he often detected elements of coarseness) observed, Bert Coote as Clown, his sister Carrie as Columbine and Connie Gilchrist as Harlequin excelled. Bert Coote clearly delivered a tour de force with his 'sneezing business':

> All these sneezes are reproduced by Master Bertie Coote, who, moreover, proves himself an adept in the art of illtreating Pantaloons, chaffing his audience, jumping through windows, robbing butchers and bakers and candlestick makers, sitting on babies, 'flooring' policemen and carrying on 'the same old game,' practised by Clowns from time immemorial.[29]

In *Little Red Riding Hood* 'Master Bertie Coote again shone conspicuously as Clown, playing his practical jokes, dancing his hornpipe, and singing his songs of "Hot Codlins" and "Tippitwitchet" as though to the manner born',[30] supported by his sister Carrie and Connie Gilchrist, graceful and accomplished as ever. With the Harlequinade running to twelve pages in the licensing copy it is clear that Blanchard did not stint in his attention to it.[31] The Harlequinade in *Robin Hood* also runs to twelve pages with the added attraction: 'To make our Children's Pantomime beat all / From other forms another troop I call' at which point 'Double Pantomime Company appear'.[32] Nevertheless it was the seasoned young performers who received the plaudits:

> In addition to general intelligence and untutored spontaneity there are two somewhat marked features of excellence in the Adelphi Children's Pantomime this year. The tricks are generally good, the fun is continuously buoyant, but the rare grace of Miss Connie Gilchrist as the chief Harlequin and the ready humour of Master Bertie Coote as the Liliputian Clown of Clowns deserve special recognition. The elegance of the steps of little Connie Gilchrist when she dances with Carrie Coote in 'La Truandaise' attracted immediate attention; and following up his great success of last year Master Bertie Coote surpassed himself in 'Hot Codlins,' taking the audience into his confidence and making his comical expression of countenance a complete study. This was not the ordinary stereotyped Clown's jingle, but a thoroughly characteristic delivery of a comic song, which tickled the gallery boys not a little. [33]

The Victorians' predilection for child performers is often regarded as suspect by later generations, but it is abundantly clear that the Children's Pantomimes at the Adelphi were characterised by the outstanding talent of their young performers.

Other attractions included[34] some topical references such as (in *Little Goody Two Shoes*) the scene in the Village School of Goody Two Shoes with her mercenary attitude ('by keeping school / I have grown richer'), her reliance on corporal punishment ('large rod before her'), the gender based curriculum ('Now, girls, go on with sewing – boys, your sums', pp. 7–8) and her eviction by 'her cruel landlord' (p. 10), who takes exception to 'May-Day mummeries' (p. 17). *Robin Hood* features the Nottingham 'Great May Fair' in 1188, the Sheriff of Nottingham's threat 'to inclose the forest',[35] a few lines lifted from Shakespeare spoken by Robin himself: 'Now, my co-mates and brothers in exile …' (p. 9), and an inebriated Friar. All three Children's Pantomimes incorporated scenes set at sunrise or twilight thereby affording the management the opportunity to show off its lighting capacity.

Chatterton's departure from the Adelphi in 1878 apparently marked the end of its Children's Pantomimes though the convention of children's companies was to continue with Gilbert and Sullivan.[36]

The Theatre Royal, Brighton

Described by his more celebrated contemporary E. L. Blanchard as 'a clever young song writer and pantomime author'[37] Frank W. Green enjoyed particular success at the Theatre Royal, Brighton, where Mrs H. Nye Chart assumed the management following the death of her husband on 18 June 1876. Pantomimes were already well established by then, the *Era* hailing *Twinkle Twinkle Little Star* for the 1875–76 season as 'Mr Chart's twenty-second Pantomime'.[38] This was naturally a tradition that Mrs Chart was as keen to maintain as she was the theatre's reputation for respectability. H. C. Porter vouched for her success on both counts:

> To her indomitable perseverance, playgoers are indebted for the continuous round of entertainments presented week after week, and the magnificent pantomime spectacles that as each festive season comes round are developed for the amusement of old and young. Mrs Chart has secured high honours as an actress, and as a lady qualified to elevate the tone of the drama intellectually I know of no one better fitted ... Following on the steps of her husband, she has raised the standard of dramatic entertainments in Brighton and on no occasion can I really remember one instance of her having failed to keep faith with the patrons of the theatre.[39]

The presence of such patrons was observed by the critic of the *Era* on Christmas Eve 1878, when *Cinderella* 'enlisted the earnest attention and appreciation of a numerous and fashionable audience'.[40] Such patrons could be assured, wrote Leopold Wagner, 'that Mrs. Nye Chart's excellent theatre at Brighton annually presents to its fashionable patrons a Pantomime little, if at all, inferior to a similar production in the West End of London'.[41] Someone who was in a position to make such a comparison was Lewis Carroll, who between January 1874 and January 1885 attended performances of ten pantomimes at the Theatre Royal, Brighton, while staying with his Christ Church contemporary the Revd Henry Alexander Barclay whose wife and two children

(Ethel Florence – born 1864 – and Noel) were ideal companions for such entertainment.[42]

Frank W. Green

Although it was in Brighton that Green's pantomimes became an annual fixture they were widely performed elsewhere, including Covent Garden, at suburban London theatres such as the Surrey and the Pavilion in Whitechapel, and in the provinces including the Alexandra Theatre in Liverpool. Like other pantomime authors Green recycled his work, updating topical references, changing local allusions and taking the opportunity to make improvements. Sometimes theatres would publish the text of the pantomime (beautifully illustrated in many cases); licensing copies were of course submitted to the Lord Chamberlain by the theatre in which that pantomime was going to be first performed, but as popular pantomimes circulated from one theatre to another it is apparent from reviews that changes, along the lines indicated above, were introduced. Therefore, although Lewis Carroll's attendance at Green's pantomimes forms the basis for the second half of this chapter, variants are also referred to so as better to establish the distinctive quality of Green's work and its appeal to audiences further afield.

The single most important consideration for Green and the audiences who patronised Green's pantomimes was their respectability, as Frederic Leader, the manager of Her Majesty's Theatre (Haymarket), proclaimed in the text of *Little Red Riding Hood*:

> To the public.
>
> In presenting LITTLE RED RIDING HOOD to the Public, the management may be pardoned for saying that their endeavour has been to provide a pretty and familiar nursery tale, told in a simple and sustaining style, combining wholesome and genuine fun, scenic display, tasteful dresses and effects worthy of the magnificent Theatre.[43]

What then did audiences look for in a Green pantomime? Naturally it had to be fun, verbally witty with that essential ingredient – puns. Topical and local references were expected, but though these would contain elements of social and even political criticism they were placed within a framework of undeviating patriotism. Spectacular scenery, pre-eminently the grand transformation scene, was *de rigeur,* but although Green scripted these it was the manager and his/ her scenic artists who realised them. For Carroll a particular attraction of Brighton pantomimes must have been the presence in the cast of such promising child performers as 'Little Lizzie Coote' with whom he was shortly to form such an attachment at the Adelphi Theatre and whom he praised as 'a charming sprite' in *Froggy would a wooing go* in January 1875[44] and whom he 'was sorry to miss' in *Twinkle Twinkle Little Star* the next year.[45] The 1882–83 offering of *Little Red Riding Hood* found not only Lizzie back in Brighton – as Little Boy Blue – but also her sister Carrie in the title role: 'Never was or will be a more pretty or charming Red Riding Hood than Carrie Coote, and to her sister Lizzie the chief honours of the night were deservedly awarded', wrote the *Era*.[46] At the Adelphi Carrie had triumphed in the Harlequinade which Blanchard made such a feature of the Children's Pantomimes there, but Green in contrast generally seems to have left that to others, for instance John Martinette in the case of *Froggy would a wooing go*: 'The comic sections have been invented and written by John Martinette, a great favourite, whose assumption of Clown has never been equalled on these boards, under whose direction the Harlequinade runs smoothly.'[47] The note on the licence copy of Green's *Dick Whittington and His Cat* for the Pavilion Theatre in Whitechapel, reads: 'Comic Scenes to follow', providing further evidence that Green did not concern himself greatly (if at all) with them.[48]

The *Era* critic remarked of *Aladdin* (1880–81) at the Theatre Royal, Brighton 'some local allusions are cleverly introduced'.[49] 'Cleverly' no doubt, but also quite easily as in *Twinkle Twinkle Little Star* (1875–76):

Scene 1.	The Icebound House of King Winter Discovered – Hurricanes, Snowball, Icicles and other Winter characters
Snowball:	Who [the Hurricane] last November rudely folks did frighten And nearly blew a ship ashore at Brighton …
King Winter:	You'll [Hurricane] raise the price of coals already dear To what it was about this time last year.[50]

The complaint about rising prices was a recurrent refrain. In the same play Dame Tucker, formerly the woman who used to live in a shoe, but now relocated to 'more commodious quarters', having rejected 'Australian beef' ('There's heaps as praises it – but no one eats it') protests:

The baker's and the butcher's bills
My poor brains sadly flumix
And so I find it hard to fill
The children's little stomachs.
(p. 14)

This preoccupation with the cost of food may be a contributory factor to the popularity of kitchens for pantomime scenes, for Green uses them in *Dick Whittington* and *Twinkle Twinkle Little Star.* Another favoured Green theme was education. The connection is made gender specific in *Froggy would a wooing go* by the Queen: 'Girls although educated and well looking / Do not make good wives if they can't do cooking.'[51] By the 1870s the first cohort of children to have benefited from the education bills taken through parliament by W. E. Forster between 1868 and 1872 were also providing audiences for pantomimes. As with Blanchard's *Little Goody Two Shoes* this provides the context for what was described as the 'May Scene' in *Twinkle Twinkle Little Star*:

Dame Tucker: Are you a visitor miss from the school board?
I keep no children home to save the fee
I'd scorn to do it!
(p. 36)

Minerva Birch, the schoolteacher in *Little Red Riding Hood* (1883–84) picks up the same theme: 'Answer all questions, never say you can't / Then you'll increase my government grant' (p. 20). Minerva Birch's name obviously carries associations of corporal punishment along the lines of:

Dame Tucker: As substitute for supper birch I'll try
I've often heard it was an excellent schooler
That's what I call our home *ruler* ...
(p. 14)

With his dexterous play on words Green had of course introduced another topical political issue: Home Rule for Ireland. This features particularly prominently in *Jack and the Beanstalk*, seen by Carroll in January 1878, but '[re-]written expressly' for the New Theatre Royal, Bristol under the management of George Macready Chute, and opening on 24 December 1880. Earlier that year Gladstone, whose political career revolved so much around Home Rule, had again been returned as

Prime Minister. More of the flavour of the piece is given by its full title: *Jack and the Beanstalk or Harlequin Fairy Kind Heart, the Pixey King, or the Wicked Squire who had his fling*:

King: What foes have I? They're quiet in the East,
And Russia's friendly, for a time at least,
Within the House of Commons no obstruction,
In Ireland now they only raise a ruction.

The chief perpetrator of Irish unrest is revealed to be Rumplestiltskin, King of Low Spirits, a bad lot that gets knocked down:

I caused those strikes that made such sad disaster
Set master against man, and man 'gainst master,
In Ireland I set all the people hooting –
About Home Rule and prompted landlord shooting.[52]

One such 'master' is the reprehensible Squire Graball, 'No greater Villain ever has been known / Since "Dick the three times" sat upon the throne'. With a 'Chorus' of Farm Labourers chanting, 'Let's boycott him, our rent we will not pay' (p. 8), it is small wonder that a stage direction states: 'Squire comes down in a very melodramatic style' (p. 11). This pantomime abounds in topical references from the Dame's: 'I'll go and change my name to Burdett Coutts / And like her, I may marry late in life' (the Baroness was to marry in 1881 aged 69), to its 'Medley on American Airs First Time in England' and its 'Girl in the Tight Fitting Jersey' ('You see I'm quite in the fashion' p. 41), but this was not to the exclusion of more traditional elements, notably alcohol and its effects. King Pixiewixie orders: 'See that the fiddlers have their dram of dew-drops / From Campeltown, and let it be a few drops'; Dame contemplates 'taking a pub' and the Giant, given to 'takin' brandy in his tea' is much the worse for wear and even gets muddled about the sequence of 'Fee fum fi fo' (p. 28). The perils of drink feature in several Green pantomimes including *Twinkle Twinkle Little Star* in which the Emperor 'sings tipsily' and Twinkle says of Dame Tucker:

It's sad to think a little drink her intellect will smother
But it's too true a glass or two is a little too much for your mother.
(p. 39)

These lines show Green's skill with the conventional rhyming couplet of pantomime; he was no less adept at the pun, a particularly charming example being in the Cloudland scene in the same play: 'a sirius affair' (p. 25). Green has the ability to gently mock the form of which he was such a gifted exponent. Thus in *Dick Whittington and his Cat,* lines about 'the Land League' and 'the Irish land bill' notwithstanding, Green's dialogue for the chance re-union of Alice and Dick is delightfully arch:

Dick: Alice.
Alice: 'Tis he!
Dick: 'Tis she.
Alice: 'Tis me.
Dick: 'Tis thee.
Both: It be – I see – 'tis us, that is – 'tis we.
(p.20)

And later 'Omnes recognise each other' (p. 32).

With some of the action set on the 'Deck of the Rolling Mary Ann' (with Alice disguised as a sailor) Green's *Dick Whittington* takes on the air of a nautical melodrama in the vein of Douglas Jerrold's *Black-Ey'd Susan*:

Dick: Tho' buffets on the land and sea may hurt me
I never let my British pluck desert me.
Song – 'Old England ne'er shall lose her sway'.
Hornpipe (p. 29) ...
Dick: As a Briton I protest 'gainst this attack
And claim protection of the Union Jack.
(p.35)

The national flag was also pressed into service in *Jack and the Beanstalk* 'GRAND NATIONAL SONG The Glorious Union Flag' (p. 43) and in *Cinderella* at Brighton in December 1878:

> In the ball-room scene are introduced representatives of Russia, Turkey, Austria, Germany, Spain, Italy, America, France, Scotland, Ireland, Wales and England, with Britannia. The last bears the standard, The Prince, Cinderella, and Britannia sing portions of 'Here stands the post', a batch of sailors supporting the chorus.[53]

Audiences at the lavish productions of Green's pantomimes got a lot for their money – child performers a-plenty, including a prodigy or two such as the young Cootes, and spectacular scenery (thirteen changes in the Harlequinade alone for *Cinderella*) – but a clear message to emerge from revisiting – albeit imperfectly – his work is that for all its humour, wit and undeniable social satire it did ultimately endorse a patriotic and cohesive view of Britain, a view which we are bound to suppose commended itself to audiences as diverse as those at the fashionable Theatre Royal Brighton and their counterparts at the Pavilion Theatre in Whitechapel. Some similar topical references are to be found in Blanchard, notably to education; both authors are prone to quote Shakespeare (Green even slips into Latin – 'ipse dixit' – occasionally) and traditional themes such as the effects of alcohol also appear, but these are less pronounced than in Green, as is the call to patriotism. In contrast the Harlequinade retains more prominence with Blanchard than with Green, in no small measure because of the talents of Connie Gilchrist and the Coote siblings at the Adelphi. Of the Cootes, who of course formed a link with Brighton, Carroll wrote to Gertrude Chataway '[they are] friends of mine, and very nice children they are – and wonderfully clever'.[54] What more could he ask for?

Notes

1. Richard Foulkes, *Lewis Carroll and the Victorian Stage: Theatricals in a Quiet Life* (Aldershot: Ashgate, 2005).
2. Clement Scott and C. Howard (eds), *The Life and Reminiscences of E. L. Blanchard, with notes from the Diary of E.L. Blanchard*, 2 volumes (London: Hutchinson, 1891), vol. 2, p. 560.
3. A. E. Wilson, *Pantomime Pageant* (London: Stanley, Paul & Co., 1946), p. 61.
4. Scott and Howard, *The Life and Reminiscences of E. L. Blanchard*, vol. 2, p. 684.
5. Edward Wakeling (ed.), *Lewis Carroll's Diaries*, 10 volumes (London: The Lewis Carroll Society: 1993–2007), vol. 5, 1999, p. 45.
6. E. L. Blanchard, 'The Playgoer's Portfolio. History of the Adelphi Theatre', *The Era Almanack, 1877* (London: *The Era*, 1877), p. 10.
7. *Era*, 30 December 1877.
8. Wakeling, *Lewis Carroll's Diaries*, vol. 6, 2001, p. 378.
9. *Era*, 27 December 1874.
10. Wakeling, *Lewis Carroll's Diaries*, vol. 7, pp. 13–14.
11. Ibid., pp. 16–17.
12. *Era*, 12 August 1877.
13. Ibid.
14. Wakeling, *Lewis Carroll's Diaries*, vol. 7, p. 76.
15. Tracy C. Davis, 'The Employment of Children in the Victorian Theatre', *New Theatre Quarterly*, 2:6 (May 1986), pp. 117–35.

16. Wakeling, *Lewis Carroll's Diaries*, vol. 7, p. 75.
17. Langford Reed, *The Life of Lewis Carroll* (London: W. & G. Foyle Ltd. 1932), p. 95.
18. Morton N. Cohen (ed.) With the assistance of Roger Lancelyn Green, *The Letters of Lewis Carroll*, 2 volumes (London: Macmillan, 1979), vol. 1, p. 371.
19. Wakeling, *Lewis Carroll's Diaries*, vol. 7, p. 29.
20. Ibid., p. 14, n. 8 in which Edward Wakeling refers to a photograph identified as being of Connie Gilchrist by Carroll in the Chicago Art Institute.
21. Horace Wyndham, *Chorus to Coronet* (London: British Technical and General Press, 1951), p. 91.
22. Wakeling, *Lewis Carroll's Diaries*, vol. 7, p. 93.
23. Cohen, *Letters of Lewis Carroll*, vol. 1, p. 295.
24. Wakeling, *Lewis Carroll's Diaries*, vol. 7, p. 98.
25. Cohen, *Letters of Lewis Carroll*, vol. 1, p. 335.
26. Wyndham, *Chorus to Coronet*, p. 88.
27. Cohen, *Letters of Lewis Carroll*, vol. 1, p. 270, n. 1.
28. Anne Varty, *Children and Theatre in Victorian Britain 'All Work, No Play'* (Basingstoke: Palgrave Macmillan, 2008), p. 27.
29. *Era*, 7 January 1877.
30. *Era*, 12 August 1877.
31. *Little Red Riding Hood or Harlequin Grandmama* by An Old Boy (London: Aubert's Steam Printing Works), printed copy submitted to the Lord Chamberlain (53189L), British Library. Date of Licence July 30 1877. For this and other pantomimes page number references are given in the text.
32. *Robin Hood and his Merry Little Men* by An Old Boy, submitted to the Lord Chamberlain (53196 H), British Library. Date of Licence 18 December 1877.
33. *Era*, 30 December 1877.
34. *Little Goody Two-Shoes or Harlequin and Little Boy Blue* by An Old Boy (London: Aubert's Steam Printing Works), printed copy submitted to the Lord Chamberlain (53178 H), British Library, date of licence 21 December 1876.
35. *Illustrated London News*, 29 December 1877.
36. Leslie Ayre, *The Gilbert and Sullivan Companion* (London: Pan Books, 1972), p. 59.
37. Scott and Howard, *The Life and Reminiscences of E. L. Blanchard*, vol. 2, p. 560.
38. *Era*, 2 January 1876.
39. H. C. Porter, *The History of the Theatre of Brighton from 1774 to 1886* (Brighton: King and Thorn, 1886), p. vii.
40. *Era*, 29 December 1876.
41. Leopold Wagner, *The Pantomimes and all about them. Their Origin, History, Preparation and Exponents* (London and Manchester: John Heywood, 1881), p. 23.
42. Wakelin, *Lewis Carroll's Diaries*, vol. 6, p. 205, n. 320.
43. Frank W. Green, *Little Red Riding Hood* (London: Fox & Co., 1883), 'To the public'.
44. Wakeling, *Lewis Carroll's Diaries*, vol. 6, p. 377.
45. Ibid., p. 444.
46. *Era*, 6 January 1883.
47. *Era*, 3 January 1875.

48. *Dick Whittington and his Cat*, submitted by the Pavilion Theatre Whitechapel to the Lord Chamberlain (53264M) in the British Library. Date of licence 20 December 1881.
49. *Era*, 8 January 1881.
50. *Twinkle Twinkle Little Star*, submitted to the Lord Chamberlain by the Theatre Royal, Brighton (53158C) in the British Library. First performance 27 December 1875.
51. *Froggy would a wooing go Harlequin Lily White Duck The Pretty Princess & the Fairies of the Dancing Water*, submitted to the Lord Chamberlain by the Theatre Royal Brighton (53145A) in the British Library, p. 19. First performance 26 December 1874.
52. *Jack and the Beanstalk or Harlequin Fairy Kind Heart, the Pixey King, or the Wicked Squire who had his fling*. Written 'expressly for … the New Theatre Royal … Bristol … 24 December 1880', for sale in the theatre, p. 6.
53. *Era*, 29 December 1878.
54. Cohen, *The Letters of Lewis Carroll*, vol. 1, p. 270.

4
Harlequin Encore: Sixty Years of the Britannia Pantomime

Janice Norwood

> Altogether, I seriously recommend those of my readers who find a pantomime once a year good for them, to go next year to the Britannia, and leave the West End to its boredoms and all the otherdoms that make it so expensively dreary.[1]

So George Bernard Shaw exhorted readers of the *Saturday Review* in April 1898. By then the pantomime at the Britannia Theatre, Hoxton, on the borders of the East End of London, had become something of an annual institution. The first was performed in 1841 and apart from the following year when the theatre was not licenced, a new production opened every Boxing Day up to 1901.[2] The Britannia therefore provides an interesting case study to explore changes in pantomime practice during the Victorian period and to identify the elements that differed from those of other London theatres.

Some conclusions are inevitably speculative since none of the Britannia's pantomimes was published and no prompt copies have survived. All save the 1841 and 1843 scripts are preserved in the Lord Chamberlain's licencing collection. We know, however, that such scripts are frequently inadequate because performers inserted ad-libs and comic business. For example, a review of *Little Busy Bee, or The Old Lady of Threadneedle Street* (*Era*, 1 January 1865) mentions 'some excellent allusions to Peabody's gift, marriage for money, the Yelverton case, [and] the German Legal Protection Society', yet none of these topical issues is obvious in the Lord Chamberlain's licensing copy.[3] Moreover, there is a marked variation in the level of detail, the Harlequinade is not included until 1851, and many of the later manuscripts fail to specify stage directions or song lyrics. Fortunately, the pantomimes were widely reviewed, and playbills and programmes provide additional information.

The reviews highlight the enormous number of theatregoers who attended the Britannia's pantomimes each year. Arguably, the genre's myriad attractions – music, dancing, comedy, agility, fantastic costumes, lavish sets, mechanical wizardry and visual spectacle – explain its continued popularity. Representations of magical kingdoms contrasted with comic treatments of reality in the Harlequinade. Giants, animals and fairies were juxtaposed with allusions to contemporary people and events. Many of the features that the audience enjoyed in other genres reappeared. Like melodrama, pantomime plots were based on the dichotomy of good and evil, and exhibited the same veneration of the honest poor and censure of the tyrannous rich. The staging, particularly the transformation and chase scenes, satisfied the taste for the spectacular that was fed by sensation drama. Moreover, because the Britannia's pantomime retained many of the conventions that other establishments discarded, it had the attraction of tradition. Most importantly, because of the dynamic interaction of the audience with the performance, the pantomime nurtured a sense of community.

The pantomime openings were written by a small number of playwrights (see Table 4.1). Until the end of the 1870s they tended to be performers at the Britannia (only Collier, Hazlewood and Merion were not).[4] In contrast, Spry, Bowyer and Addison, who between them wrote all the pantomimes between 1880 and 1899, were dramatists, not actors, and did not produce any other work for the Britannia. Seven of the early pantomimes have an unidentified author.

The standard of writing displayed in the scripts is uneven. The best, mainly written in rhymed verse, are peppered with puns and verbal jokes. A typical example appears in Frederick Bowyer's *Queen Dodo, or Harlequin Babilo and the Three Wonders* (1883). Babilo, the hero, kisses Princess Merryheart, who has been turned into a statue. She wakes from the spell and sneezes, to which he says, punning on 'a-tishoo': 'She knows me then[.] I swear she said tis you.'[5]

The plot of the 1851 pantomime, *Harlequin and the Koh-i-noor, or the Princess and the Pearl* is representative. It was written by Charles Rice, a comic actor then engaged at the Britannia. Two of his melodramas (*The Merchant and the Mendicant* and *The Man of the Red Mansion, a Tale of the Hundred Days*) had been produced at the theatre earlier in the year. His portrayal of the beggar in the former was lauded as 'a natural piece of acting' (*Theatrical Journal*, 3 September 1851). Yet a later article in the same journal (8 April 1857) decried his talent: 'He is many degrees

Table 4.1 Authors of Britannia pantomime openings

Dramatist	Date of pantomimes	Total
John ADDISON	1887–1899	13
Colin HAZLEWOOD	1856–57 (with Seaman), 1859–60, 1863–67, 1873–74	11
Frederick MARCHANT	1861–62, 1868–71, 1875–1878	10
George Dibdin PITT	1844–45, 1847–48	4
Frederick BOWYER	1883–86	4
William SEAMAN	1856–57 (with Hazlewood), 1858	3
H. SPRY	1880–82	3
Charles WILMOTT	1900–01	2
Charles RICE	1851	1
J. W. COLLIER	1852	1
William ROGERS	1854	1
Charles MERION	1872	1
J. B. JOHNSTONE	1879	1
Unknown	1841, 1843, 1846, 1849, 1850, 1852, 1855	7

below the worst comedian I know of for vulgarity, liberty, and interpolation; he is a clever man, probably, but he has an odd way of making his cleverosity [*sic*] palpable.' *Harlequin and the Koh-i-noor* suggests Rice's talent as a writer was superior to this evaluation and indeed in the 1870s he became actor-manager at Covent Garden Theatre and wrote four of its pantomimes.[6]

Rice's Britannia opening consists of five scenes before the main characters are transformed into the Harlequinade figures. The names of the characters are indicative, including the slave driver, Yooz-a-Khat (as in 'Use-a-cat-o'-nine-tails', referring to the fearsome whip) and Kohm-i-cal Fellah, the Captain of Guards. The use of such names was a recognisable feature of the Britannia's pantomime house style. Hence in October 1869 the stage manager Frederick Wilton wrote a letter 'to Marchant, author of forthcoming Pantomime, complaining of there being no punning descriptions of characters in his opening'.[7]

The plot centres on the Princess Heermee-Singh, daughter of Killallee Khan, the King. She loves Ameer Yooth, but her father has forbidden the match because of his impecuniosity. The King decrees that whoever can bring him the biggest diamond shall marry the Princess. Ameer, assisted by the Spirit of Night, finds the Koh-i-noor, but is warned that it will not sparkle in a land ruled by a despot. The scheming slave driver seizes the diamond and claims the Princess. The King, ignoring the entreaties of his daughter, consents. Discovering that the diamond has turned black, the King dismisses Yooz-a-Khat as a swindler and promises to let Ameer marry his daughter if he can produce another gem. The Princess bravely swims to a grotto beneath the ocean to beg the Queen of the oysters, Mother o' Pearl, for a large pearl. Koortah, an old oyster, offers the Princess that of his dead wife if she will marry him. He attempts to force himself on her, but she escapes and presents the pearl to her father. The Spirit of Night then transforms the stage into a brilliant cavern of jewels. Koortah appears to reclaim the pearl, so the King, alarmed at losing it, coldly agrees to his taking the Princess in exchange. Just then Mother o' Pearl arrives and changes the lovers into Harlequin and Columbine, the tyrant King into Clown, and Koortah into Pantaloon.

The pantomime called for elaborate staging. One scene, representing the 'Bower of Fuchsias', was praised in the *Theatrical Journal* (7 January 1852) as 'one of the most gorgeous scenic effects we ever witnessed'. Rice's script gives detailed stage directions. For Scene 3, which takes place in a corridor of the palace, they call for 'a row of arches along the flats – each arch springing from a huge stone Head of comical expression'.[8] These echo the big heads that the main characters traditionally wear in a pantomime opening. When the princess is dragged on by Yooz and threatened with rape, the ladies of the court weep. The stage directions state:

> *As they cry the large faces on the wall cry also, large tears an inch in size rolling down their cheeks into their mouths – and Men with speaking trumpets roaring the chorus out of their mouths from behind Flats.*[9]

At other strategic moments, the heads laugh.[10] They thus function as a chorus to the action.

In a later scene Rice skilfully exploits the comedic potential of the characters' big heads. Ameer and Yooz are confined in a prison cell. After a fist fight that leaves Yooz insensible, Ameer uses his adversary's unfurled turban as a rope ladder to aid his escape out of a high window. On regaining consciousness, Yooz tries to follow:

> *[He] puts his head out as [a]bout to escape, when Ameer closes down the portcullis so holding Yooz's head outside & his body in, he calls out murder & gradually lowers his body, his neck stretching six feet[.] The guards run on hearing his cries & seeing him escaping, they catch hold of his legs which they stretch in like manner, at last his body falls leaving his*

> *head & his legs are pull'd off – The King enters, starts at seeing what has happened, his mask has assumed an appearance of grief[.]*[11]

This physical humour prefigures the fun of the Harlequinade, known at the Britannia as the comic scenes. As is generally the case, the four comic scenes are set in contemporary urban locations.[12] The first features the interior of Clown's Exhibition Room, a topical reference to the Great Exhibition held at the Crystal Palace that year. This is particularly apt since one of the attractions of the exhibition was the display of the Koh-i-Noor diamond, which had recently been passed to Queen Victoria after Lahore became part of the British Empire. During the scene two chests are carried on stage, one bearing the legend 'Prize Method of feeding Paupers, on the most Economical Principal', and the other marked 'Best Treatment for Convicts'. Harlequin reveals a badly clothed workhouse girl and a 'fine fat healthy convict' in prison dress. Clown and Pantaloon kick the convict off stage and present the girl with a leg of mutton and money. Thus although much of the action seems anarchic, with the Clown frequently stealing, tricking or attacking people, the scene conforms to the usual Britannia preference of rewarding the deserving poor and punishing criminals. The other scenes are set in 'A Snugg [*sic*] Apartment' outside a chandler's shop and a bloomer emporium, and finally in an equestrian tavern and cook shop.

Charles Dickens made a useful comment about the Britannia's comic scenes after attending the 1859 pantomime, *Spirit of Liberty, or Harlequin Needles and Pins and Europe, Asia, Africa and America*. He contrasts its realistic portrayal of character types with those that featured in the opening:

> I noticed that the people who kept the shops, and who represented the passengers in the thoroughfares and so forth, had no conventionality in them, but were unusually like the real thing – from which I infer that you may take that audience in (if you wish to) concerning Knights and Ladies, Fairies, Angels, or such like, but that they are not to be done as to anything in the streets. I noticed, also, that when two young men, dressed in exact imitation of the eel-and-sausage-cravated portion of the audience, were chased by policemen, and, finding themselves in danger of being caught, dropped so suddenly as to oblige the policemen to tumble over them, there was great rejoicing among the caps – as though it were a delicate reference to something they had heard of before.[13]

This identification of the audience with characters evading the forces of law is an example of what David Mayer terms 'retributive comedy'.[14] Chiefly occurring in the Harlequinade, such episodes enact the desire to overthrow or outwit the dominant power. The appeal of these scenes is similar to that of the puppet Punch (himself the subject of the 1853 Britannia pantomime) where satire and slapstick are used to criticise authority.

During the course of the nineteenth century other theatres gradually reduced the Harlequinade until it virtually disappeared. At a time when the perception of the general deterioration of the pantomime was frequently expressed, the Britannia was recognised as one of the last institutions to retain the old tradition.[15] For example, in 1881 Leopold Wagner wrote:

> we have still in existence a people's theatre, in the East-end of London, whose patrons are annually treated to a Pantomime of the old sort, quite different from anything to be witnessed at the other houses. Indeed, anyone who has made a study of English Dramatic Literature would not fail to recognise, in a visit to the Britannia Theatre, as near an approach to the old Morality Plays as might be expected in the present generation.[16]

The Britannia made a virtue of the traditional format; for example, trumpeting the 'Good Old-Fashioned Comic Scenes by George Lupino and Troupe' in advertisements for the 1894 pantomime *The Giant of the Mountains* (*Era*, 9 February 1895).

Another distinguishing feature of the Britannia's pantomimes is the originality of the plot. (Hazlewood's 1873 pantomime, *Cocorico, or the Hen with the Golden Eggs*, which translated a French tale, was exceptional.) During the 1850s and 1860s some of the subjects were derived from nursery rhymes, such as *Hush-a-by-Baby* (1856) and *Hickory Dickory Dock* (1863).[17] Others took inspiration from historical or literary characters, such as Cleopatra in *Egypt* (1854), *Abon Hassan* (1862), which was vaguely based on the Arabian nights, and *Don Quixote* (1867). The 1848 and 1855 pantomimes were inspired by the legendary figures of Old Parr, who lived to 152 years and is buried in Westminster Abbey, and the prophetess Mother Shipton. Although these pantomimes drew on popular culture for the subject matter, the plots developed new story lines. By the 1880s all the Britannia's pantomimes were original inventions. A report on the 1883–1884 season (*Era* 19 January 1884), calculated that there were only 23 subjects at the 104 theatres producing pantomimes across the county. *King Aboulifar*[18] and the Britannia's *Queen Dodo* were the only original contributions. Novelty was an astute marketing ploy; customers knew they could not see a better version of the Britannia's pantomime elsewhere. Had the theatre produced versions of *Aladdin, Puss in Boots* and the like, it would have been in more direct competition with neighbouring theatres.

Although the plots and characters seem fantastical, some, such as Hazlewood's *Spirit of Liberty*, were grounded in the experience of the predominantly working-class audience. Its inventive plot concerns the struggle between a race of Needles (supported by Liberty) and their adversaries, the Pins (aided by Atropos, the Goddess of Discord). The industry of the Needles is contrasted with the idle Pins, who are only fit for mending the garments made by the Needles. Following an unsuccessful assault by the Pin Army, the Needles are assailed by Rusty Fusty, the corroding sprite, and magnetically drawn towards the Loadstone Rock, where their fate is to become the slaves of the Pins. The audience, many of whom would have worked in the local tailoring and sewing trades, is intended to empathise with the Needles. The King of the Needles makes reference to the low earnings of seamstresses: 'The Needles the great instrument of the clothing trade/ And may those who live by needlework get much better paid.'[19] This identification is underlined by the fact that the Needles are championed by Liberty, played by actor-manager Sara Lane, who is nominally the Spirit of Europe, but more specifically represents England. Negating Janice Carlisle's claim that 'Christmas entertainments erased all distinctions of race, class and gender in order to emphasize the homogeneity of British patriarchal rule', the implication is that it is the honest industry of the East Enders that makes the country great.[20]

Inventive plots do not, however, appear to have been the audience's main interest. Contemporary evidence suggests that the crowds at Boxing Day performances (to which so many of the reviews refer) were particularly boisterous. The level of noise prohibited hearing much dialogue, as shown in this evocative description of the Boxing Night audience for *Hickory Dickory Dock The Mouse that Run Up the Clock* in 1863:

> They whistled like a thousand locomotives, in a state of shrill expostulations, and as for the usual hum of conversation prevalent among a more subdued class of audience, the sound of many tongues from the dense mass of closely packed humanity in the Britannia was more like an incoming tide on a shingle beach than anything else.[21]

Hence, in his diary Frederick Wilton notes on 28 December 1864, the fourth time the pantomime was played: 'Great House again, but quiet. The words of the Pantomime listened to by the Audience for the first time.'[22]

The audience's animated response was not unique to the Britannia, but according to the *Saturday Programme* (29 April 1876) was 'far more hearty, if far more noisy and rough, than is boxing-night even at Drury Lane itself'. It was not that the good-humoured audience was disruptive, rather that it actively participated in the proceedings from the moment the orchestra began playing and joined in the musical numbers. Reviewing *Rominagrobis, the Tail of a Cat*, the *Daily Telegraph* (17 December 1877) noted:

> Mr. Lepyeat, who conducts the large orchestra must decidedly share in the general honours achieved yesterday. It is one thing to have thousands of juvenile spectators with music in their souls, or rather on their lips and another to bring out that latent vocal power, as it was brought out at least half-a-dozen times in the course of 'Rominagrobis.' There was literally a crash of sound when now and again the chords of some familiar chorus were played ...

The audience also responded to topical allusions. Although contemporary issues were sometimes addressed in the Britannia's melodramas, the pantomime genre provided more licence to satirise personalities and events. For example, in *Old Daddy Long Legs, or the Race for the Golden Apples* (1865) Sir Regent Circus and his valet, Little Britain, are blown up on 'a Metropolitan railway' and materialise in the Region of Clouds. The pantomime appeared in the year that the Metropolitan Line was extended to run into Moorgate and a new railway line, the North London Extension, started near the Britannia, events that would have interested local theatregoers. Similarly, the 1898 pantomime *King Klondike* takes its inspiration from the Klondike gold rush then taking place in the Yukon.

The comic scenes of *Mother Shipton's Prophecy of Seven Women to One Man, or Don Giovanni and the Witch's Broom* display a more savage commentary on contemporary politics (*Era*, 30 December 1855):

> Two scenes deserve special notice; one, the Ruins of Sebastopol, where Clown gets most unmercifully shelled; and another, the Rival Ratcatchers of Westminster. Here the peace-at-any-price party, the ticket-of-leave officers from the Crimea, the Aberdeen Government,

> and other popular subjects connected with the war, are hit off with great severity; one of the changes exhibited Aberdeen, an Old Rat, hanging. A Baby Show, with a prize for the most illused [*sic*] baby in England, was awarded to the baby Charles Napier; whilst to Johnny Russell the prize was given for the most deceitful and changeable. These hits were highly relished by the audience.

Here, the hanging rat represents Lord Aberdeen, whose government was defeated in 1855 partly as a reaction to the bad organisation and poor conditions endured by British soldiers during the Crimean War. Lord John Russell had been Prime Minster from 1846 to 1852, but had temporary retired from politics that year. The 'illused baby' refers to Vice-Admiral Charles Napier, who became a government scapegoat for failures in the Baltic campaign.[23]

The pantomime openings also included burlesques of other dramatic genres and performers. For example, in *El Flambo, or the Waters of the Singing Well* (*Era,* 2 January 1876): 'Mr Frederick Marchant, made up as Sir Rummefordde, was an exceedingly comic personage, always seeking in accents that recalled Mr Irving, Mr Bandmann, and other popular actors, "his lost cheoild."' Likewise, for his role in *The King o' the Castle* (1894), George Lupino was made up to suggest Edmund Kean as Richard III. It was not only individuals who were sent up: *The Man in the Moon* (1892) featured a scene aping the cave incident in Boucicault's *The Colleen Bawn; Old Daddy Long Legs* gave 'an admirable burlesque on the casino style of dancing' (*Era*, 31 December 1865); and *Queen Dodo* lampooned the musical style of the Salvation Army. It is through such references to diverse aspects of contemporary culture that the pantomime derived its power and freshness.

The dialogue suggests a significant proportion of the audience were knowledgeable about a range of contemporary issues. For example, *The Magic Dragon of the Demon Dell, or the Search for the Mystic Thyme* in 1888 contains an exchange alluding to the controversy stirred up by Mona Caird's article 'Marriage', which appeared in the *Westminster Review* and claimed wedlock was a form of sexual slavery. The pantomime dialogue also refers to the *Daily Telegraph*'s subsequent request to readers to respond to the question 'Is Marriage a Failure?'[24]

In this case the humour is linguistic, but audiences also relished physical comedy in the comic scenes. An early example comes from the 1853 pantomime, *Punch and Judy, or Harlequin Shallabalah and the Dog 'Toby'*. The *Sunday Times* (1 January 1854) describes a scene set in 'The Parlour of a Sporting House', where a man enters carrying a gun:

> The Clown (Boorn), to prevent an accident, takes the warlike implement from the sportsman, and whilst explaining (*à la* the Bishop of Bond-street) the cause of accidents, it goes off, and, to the consternation of the audience, the head of one of the box visitors falls on the stage. Presently the unfortunate victim thrusts his bloody neck forward, and, amid the exclamations and convulsive laughter of all, throws out a rope of handkerchiefs to recover his head. The trick was so well managed that even the persons sitting in the same box were not aware of the deception.

Clearly the individual talent of the Clown was paramount to the pantomime's success. The Britannia engaged a succession of skilful exponents, including Jean Louis, who played Clown in 16 pantomimes from 1854 until 1869. He was described as 'a very little fellow, who is funny without vulgarity, and who is most deservedly a great favourite with the audience' (*Penny Illustrated Weekly News,* 5 January 1867). The 'Great Little Huline', who appeared in the 1871 and 1873 pantomimes, fashioned his style on Grimaldi. The *Era* (28 January 1872) noted: 'Besides being a merry, quick Clown of the old-fashioned red-hot poker, hot codlins, hornpipe school, the Great Little Huline displays remarkable versatility by performing numerous feats of a peculiar, difficult, and pleasing kind.' After Louis, the longest-serving Clown was Tom Lovell, who appeared in ten pantomimes from 1878.

While music and witty dialogue delighted the audience's aural sense, spectacle provided visual entertainment. Each pantomime required numerous changes of backdrop and scenery. The opening typically featured supernatural or fairytale worlds alongside scenes depicting every kind of earthly, subterranean and underwater landscape. In contrast, the comic scenes called for representations of contemporary London.

Costumes supplied an extra element of spectacle. Many early reviews praised the beautiful costumes and make-up, and it was not unusual for Sara Lane to appear in six outfits during the course of one performance. Many of the pantomimes required elaborate animal or insect costumes, such as the fish and shellfish in *Harlequin and the Koh-i-noor,* the beetles in *Old Daddy Long Legs,* the cats in *Hickory Dickory Dock,* and the birds in *The Goblin Bat, or Harlequin Meloda and the Little Oof Bird* (1887).

Playbills attest to the importance of the transformation scenes. These required much technical proficiency. For Hazlewood's *Little Busy Bee* in 1864, which included a waterfall of real water, the scene gradually unfolded during the course of 15 minutes and there were eight mechanical changes for the stage hands to effect. The visual impact of these moving pictures was dependent on skilful lighting, as shown by the closing scene of *Spirit of Liberty*. The *Era* (1 January 1860) described 'the Submarine Grotto of Phosphorescent Light, or Elysium of the Naiades':

> Coral grottoes are at first seen covering the whole stage; a slight undulating motion takes place, and they gradually open and discover the Naiades lying about in shells, and from the back a large globe, about one-third the height of the stage, comes forward, and upon and around it are females in silver tissue robes. This globe is entirely composed of cut glass drops, and revolves. A similar revolving one is in the inside and through the centre passes a column of gas jets. The scene, when lit from the front of the stage with the Electric light, has a most gorgeous and dazzling effect.

The transformation scenes created a crescendo of visual excitement by introducing movement into beautiful static tableaux. Many feature the entrance of the principal fairy, as in this example from *Old Daddy Longlegs* (*Era*, 31 December 1865):

> The first part of this elaborately constructed scene represents a screen of pink bell-shaped flowers, from which rises a fairy standing on an enormous gilt butterfly. Both the insect and the lady soar gradually upwards, and disclose a lake dotted with waterlilies, and on the surface floats Queen Brilliant, in a car drawn by swans. A large framework rises at the back, and under this arched pavilion Fairies, in dresses of rich and glittering material are placed.

From 1896 to 1898 there was also a spectacular 'flying ballet' in which the fairies seemed to float through the air.

Equally important were the chase scenes. George Lupino, who was responsible for creating the complicated chases through the various traps in the pantomimes of the 1890s, was a talented acrobat. When George Bernard Shaw saw him in *The Will o' the Wisp* (1897), he was impressed not only with Lupino's agility and skill, but also with the manner of the audience's response to it:

> When a white statue which had stood for thirteen minutes in the middle of the stage turned out to be Mr Lupino, who forthwith put on a classic plasticity, and in a series of rapid poses claimed popular respect for 'the antique,' it was eagerly accorded; and his demon conflict with the powers of evil, involving a desperate broadsword

> combat, and the most prodigious plunges into the earth and projections therefrom by volcanic traps as aforesaid, was conducted with all the tragic dignity of Richard III and received in the true Aristotelean spirit by the audience.[25]

The successful production of Lupino's chase scenes testifies to the mechanical expertise of the Britannia's backstage staff. They were clearly adept at working the various traps by which performers entered or left the stage. An illustration of the 1876 pantomime, *Turlututu, or The Three Enchanted Hats*, written by Frederick Marchant, depicts one in action as Pollie Randall, playing Il Diavolo, rises through the vampire trap (Figure 4.1). The illustration, which appeared in the *Illustrated Sporting and Dramatic News* (6 January 1877), reveals several features characteristic of nineteenth-century pantomime. Around the main characters, the supporting figures wear oversized pantomime heads and ballet girls, dressed as flowers, are artfully arranged as though in plant pots.

As with Lupino, much of the popularity of the Britannia pantomime can be attributed to the strength of the casting. Sara Lane, whose singing, dancing and comic skills were ideally showcased in the pantomime, was an acknowledged crowd-puller. (She is depicted pointing in Figure 4.1.)

Figure 4.1 'Scene from *Turlututu* at the Britannia Theatre', *Illustrated Sporting and Dramatic News* (6 January 1877) (collection of Janice Norwood)

Her earliest pantomime parts were in breeches roles, such as Peter Piper in *Harlequin Bluebottle, or the Owl and Fairy Queen of the Butterfly Bower* (1846), but over the decades she played a variety of princes, princesses, fairies, servants, gnomes, sorceresses and demons. The last pantomime in which she starred was *The Old Bogie of the Sea* (1892), in which she played Baroness Awlforgood.

Each year specially engaged artists, such as Lupino, added their talents to those of the regular company. In addition, many supernumeraries were hired, although not on a scale to compete with the three or four hundred performers that Augustus Harris choreographed in spectacular processions for the Drury Lane pantomimes of the 1880s and 1890s.[26] Following the practice that became widespread during the second half of the nineteenth century, the Britannia also employed music-hall entertainers.[27] Notably, Marie Lloyd was engaged as Princess Kristina in *The Magic Dragon of the Demon Dell* in 1888.[28] The part is interesting as it plays with the music-hall favourite's identity. She first appears as a traditional chaste lover, but when an unsuccessful suitor steals a magic lamp representing her modesty, she becomes sexually predatory, chasing after men until the lamp is rediscovered in the final scene. The role was particularly apt for a star whose act was known to be risqué. Here the audience is overtly encouraged to recognise simultaneously both the performer and the role. This 'phenomenological doubling', to use Watson's phrase, similarly occurs when Sara Lane appears as herself in the 1854 pantomime *Egypt* and in *King Kookoo* (1884) where she plays the theatrical manageress Thespiana.[29] These instances illustrate how in laughing at the 'in joke', the knowing audience colludes in the artifice of the performance. Pantomime is a supremely self-conscious form of theatre.

The comic performer Chirgwin, known as the 'White-eyed Kaffir', appeared in six Britannia pantomimes. His performances included dancing, falsetto singing and playing a variety of traditional and improvised instruments, as well as hilarious ad-libs. Other well-known artists employed by the Britannia included G. H. Macdermott, Marie Kendall, Ada Reeve, Charles Coburn, Little Elsie, and Topsy Sinden. Shaw was particularly impressed with Sinden's performance as Sylvia in *The Will o' the Wisp*:

> I was agreeably astonished by Miss Topsy Sinden's dancing. Thitherto it had been my miserable fate to see her come on, late in the second act of some unspeakably dreary inanity at the West End, to interpolate a 'skirt dance,' and spin out the unendurable by the intolerable . . .

> At the Britannia Miss Sinden really danced, acted, and turned out quite a charming person.[30]

The engagement of popular performers was just one of the ways in which the theatre's management sought to provide fresh interest in what was undoubtedly the single most important production of the year. Although the pantomime demanded a substantial investment of resources, a box-office hit meant financial security for the rest of the year. The Britannia pantomime earned almost universal critical praise and in the later decades was championed as a bastion of traditional entertainment. Yet the Harlequinade is a prominent example of the potentially anarchic and it must therefore be significant that it continued to feature at the Britannia after other theatres had omitted it from their productions. The association with Sara Lane, who regularly appeared when she had virtually abandoned other performances, also contributed to its popularity. She epitomised the essence of wholesome fun that was the core of the pantomime's success. Its lampooning of the contemporary world provided the audience with a means of laughing at, and thereby temporarily negating, the stresses and challenges of life. Above all else, it was entertaining: the Britannia pantomime unfailingly provided a good night out.

Notes

1. 'The Drama in Hoxton', *Saturday Review*, 9 April 1898; reprinted in G. B. Shaw, *Our Theatres in the Nineties* (London: Constable, 1954), vol. 3, p. 355.
2. The last of the regular pantomimes was in 1901, although *The Goblin of the Sea* appeared in 1904 and *Beauty and the Beast* in 1912.
3. ADD.MS 53038 C, British Library.
4. Colin Henry Hazlewood was one of the Britannia's most prolific dramatists, contributing numerous melodramas and burlesques as well as pantomimes; see Janice Norwood, 'C. H. Hazlewood (1819–1875)' in Angela Courtney (ed.) *Dictionary of Literary Biography* vol. 344 *Nineteenth-Century British Dramatists* (Detroit: Gale Cengage, 2008) pp. 163–75. Much less is known about Collier and Merion. J. W. Collier wrote at least five other plays and acted at Sadler's Wells and the Lyceum in the 1840s. Charles Merion wrote pantomimes for the Pavilion, Victoria, Elephant and Castle, Marylebone and Gaiety Theatres as well as sketches for the music hall performer Fred Williams. According to an obituary for Williams, Merion was still alive in 1916 (*Era*, 21 June 1916).
5. ADD.MS 53306 B, f.20, British Library.
6. *Red Riding Hood* (1873), *The Babes in the Wood, or, The Big Bed of Ware* (1874), *Cinderella* (1875) and *Puss in Boots* (1877).

7. Jim Davis (ed.), *The Britannia Diaries 1863–1875: Selections from the Diaries of Frederick C. Wilton* (London: The Society for Theatre Research, 1992), p. 165.
8. ADD.MS 43038 f.357, British Library.
9. ADD.MS 43038 f.357.
10. ADD.MS 43038, ff.358, 360.
11. ADD.MS 43038, ff.371–72.
12. ADD.MS 43038, ff.378–85.
13. Charles Dickens, 'The Uncommercial Traveller', *All the Year Round* 44 (25 February 1860), 418.
14. David Mayer, *Harlequin in His Element: The English Pantomime, 1806–1836* (Cambridge, Mass.: Harvard University Press, 1969) pp. 52, 56.
15. For example, W. Yardley complained that the 'music-hall element is rapidly driving from our English stage the art of pantomime' (*Theatre*, 1 February 1881).
16. Leopold Wagner, *The Pantomimes and All About Them: Their Origin, History, Preparation and Exponents* (London: John Heywood, 1881), p. 28.
17. Some titles are abbreviated. For example, playbills give the full title as *Hush-a-by-Baby upon the Tree Top, or the Comet of 1.8.5.6. without his Tail.*
18. I have been unable to trace this pantomime.
19. ADD.MS 52987 S, f.7, British Library.
20. Janice Carlisle, 'Spectacle as Government, Dickens and the Working-Class Audience' in Sue-Ellen Case and Janelle Reinelt, eds., *The Performance of Power: Theatrical Discourse and Politics* (Iowa City: University of Iowa Press, 1991), p. 173.
21. Unattributed cutting in Theatre Cuttings 65, British Library.
22. Davis, *The Britannia Diaries 1863–1875*, p. 85.
23. Andrew Lambal, 'Napier, Sir Charles (1786–1860)', *Oxford Dictionary of National Biography* (Oxford University Press, 2004); online edition, 2008 http://www.oxforddnb.com/view/article/19747 [accessed 2 June 2008].
24. See Lyn Pykett, 'Women writing women: nineteenth-century representations of gender and sexuality' in Joanne Shattock, ed., *Women and Literature in Britain 1800–1900* (Cambridge: Cambridge University Press, 2001), p. 92.
25. Shaw, 'The Drama in Hoxton', p. 355.
26. Mayer, *Harlequin in His Element*, p. 324.
27. James Lawrence Graydon, the manager of the Middlesex Music Hall, reported to the 1892 Select Committee that over a thousand music-hall artists had played in pantomimes at London and provincial theatres during the 1891/92 season, Q.2895, Report from the Select Committee on Theatres and Places of Entertainment (House of Commons Papers, Session 1892, 240).
28. Called Princess Mazina in the script, ADD.MS 53418 C, British Library.
29. Ian Watson, 'Naming the Frame: the Role of the Pre-Interpretive in Theatrical Reception', *New Theatre Quarterly* 13:50 (May 1997) pp. 161–170; 169.
30. Shaw, 'The Drama in Hoxton', p. 354.

Part 2

Pantomime, Representation and Ideology

5
Pantomime and the Experienced Young Fellow

Jacky Bratton

The starting point for this chapter is a cartoon from *Punch*, 25 February 1860 (Figure 5.1). The caption reads 'EXPERIENCED YOUNG FELLOW: *Ah, Clara, you should have seen the* Pantomines [*sic*] *that I've seen; these modern affairs ain't half so good.*' It shows middle-class children alone in a drawing-room, the little girl slumped in a chair in an attitude she would certainly not have been allowed if adults were present; her brother stands and pontificates to her, in parody of his elders. In July of the same year the artist who drew this apparently produced another cartoon showing two little boys leaning on a grand piano, in which cousin Jack, in an Eton suit, asks cousin Henry, in a pinafore, if he likes grammar, and gets the reply that Henry doesn't know because he's never tasted it. Both drawings rely on fond observation of childish experiments in status play, their jockeying for power through the assertion of greater experience. In the February cartoon the game is observed to begin when the players are even further down the hierarchy: the experienced young fellow is himself still in skirts, and can only assert his superiority to a little girl. He cannot even pronounce 'pantomime' properly. But an important part of this joke is that the child has picked up the central adult attitude to the subject about which he is instructing his sister. The observation that pantomime 'ain't half so good' as it used to be has accompanied it down the ages, from its arrival in Britain to today. Derek Salberg's 1981 book about pantomime has a whole chapter entitled 'Pantomime is never as good as it was'. He has collected statements to this effect from *The Drama* of 1822 down, via Planché, Dickens and Bernard Shaw to the *Daily Mail* headline 'It looks like curtains for the panto' in the year he himself was writing.[1] A recent book about pantomime, by Millie Taylor in 2007, confirms that indeed in the 1980s panto ceased to be a West End speciality – but that it then

Figure 5.1 'EXPERIENCED YOUNG FELLOW' from *Punch*, 25 February 1860

simply passed into the new form of touring productions mounted by a small number of large commercial providers, since, as she says, pantomime has always actually been a strictly commercial product, moulded by market forces.[2]

The growing academic interest in the form has tended to find this a difficult or indeed unpalatable aspect of its history. In November 2003 the British Society for Dance Research held a day conference at the Laban Centre, to frame the visit of a troupe of pantomimists from the Tivoli Gardens. To their surprise, registration for the event was slow. In the opinion of the organiser Peter Bassett, it was as if people still thought of pantomime as something rather cheap and nasty put on for children at the local theatre at Christmas. He had hoped that there would be much scholarly excitement about the fact that we were to see a performance in an unbroken tradition reaching right back into the eighteenth century, by a troupe that had performed the story of Harlequin and Columbine uninterruptedly, and in the open air, for more than 200 years: a figure of breathtaking theatrical authenticity. Like the remarks of the Experienced Young Fellow in 1860, this contradictory pair of twenty-first-century perceptions by the scholars and practitioners of the art of dance exemplify and underline an important trope in the theorising of popular entertainment: that it is not what it was, and we only tolerate it for the sake of our children, in whose education it holds a special place – so long as it is preserved from innovation and pollution.

In a taxonomy of performance, pantomime may be classified as the archetypal popular phenomenon, as Morag Schiach defines that discourse.[3] A popular art is a form understood to belong to *them,* rather than to *us*: theatre for the non-theatre-goer, it is childish and/or vulgar; bound by convention, on the one hand, and pestered by novelty, on the other, it is an antitype of high art, whose continued existence is liable either to be marvelled at as an anthropological curiosity, or to be regretted as evidence of the failure of education. It is the subject both of distaste and of nostalgia: for the latter part of its existence, most of two centuries, we have expected that at any moment it will be extinguished because it is in disastrous decline, always inferior to its previous manifestations. Its present form is always-already debased; its past, alone, is worth our regret as part of the world we have – always – just lost. This has been the case with pantomime in England ever since the meretricious performances of John Rich provoked Alexander Pope's savagery in *The Dunciad,* but then, after his death, the actor was regretted and immortalised as Lun, the Harlequin par excellence. According to the conventional account, voiced in Maurice Willson Disher's most self-important work, *Clowns and Pantomimes,* 'this strange old man with uncouth manners and coarse yet pleasant face had invented the Harlequin of fantasy',[4] and his pantomimic gifts made a miracle of such things as his mime of hatching from an egg warmed in the heat of the stage sun.

Throughout the eighteenth century the conflict between the legitimate stage and 'harlequin' was to do with cultural gate-keeping: pantomime courted disapproval as a vulgar importation by businessmen, to the detriment of art and of the national drama, and it was eventually challenged by the native genius of Garrick. Harlequin was then represented as Shakespeare's rival and opponent in the battle for possession of the stage, and of audience hearts and minds. Pantomime, with its dumb shows, slapstick, its purely conventional stories embellished by ever-more-fantastic decorations and transformations of the stage, was the figure for the opposition to all that the classic drama, as represented by Shakespeare via Garrick his publicity agent, stood for in terms of moral worth and artistic value. Pantomime was drama without words or, crucially, drama without the dramatist, made up of the skills of the actor, dancer, musician, comedian, scene-painter and machinist. It was arranged rather than written; by the 1790s, hired stage managers might have the creation of one or more pantomimes per year written into their contracts. And 'the dramatist', thus excluded, is championed by critical writers, and figured by them in the shape of the unchallengeable stage genius of the pen, William Shakespeare.[5]

But the nineteenth-century history is rather different: pantomime developed an artistic value of its own, a place in the hegemonic negotiation of cultural capital. Regency pantomime threw up a genius on its own account, Joseph Grimaldi, who became figured as the greatest of popular entertainers, arriving at profundity and importance through humour, romantic alterity and rebellion against the outmoded styles and moral imperatives of the drama. David Mayer, in his magisterial account,[6] shows how the comic pantomime of the early nineteenth century was transformed into a satirical/topical weapon; it was, he demonstrates, the only satire on the stage of the day, achieving distinction largely through the work of Joseph Grimaldi and the arrangers who worked with him. And then, after the political edge of his work ceased to be important, Grimaldi and his early death continued to provide a touchstone for nostalgia: Joey the clown, tumbling in pain through the hoops of the Harlequinade, laughing through his tears, singing 'Tippetywitchet' as if it were 'On with the motley', a sentimental construct that has dominated thinking about clowning to this day.[7]

Ever since Joe died, this story goes, the pantomime Harlequinade has been dying too: throughout the nineteenth century the story of the Harlequinade in England is a story of decline, of the steady swamping-out of the transformation, the lovers, the chase, the magic bat and the greedy clown by increasing size and spectacle and imported music hall acts – as if there were some pure essence there that is for ever being diluted. The *Era* (20 February 1859) published one of many impassioned disquisitions on the subject, this time from 'W.D.' of Holloway, who blustered:

> I take the liberty of informing you that some fifty years ago a certain actor, named Grimaldi ... took to performing Pantomime Clown, and made such a hit in the part that none of his competitors could come near him. Now, why don't some of you try and do the same? ... when Pantomimes were in their zenith, and had first-rate Pantomimists to support them, that line of business was not then looked down upon with such vulgar contempt by the profession as it is now ... The harlequinade portion of each Christmas Pantomime is gradually dwindling down into an unwelcome incumbrance.

Hence the marvelling excitement at the Laban Centre about the idea that the Tivoli performers would bring with them that essence of Georgian pantomime, unpolluted by the subsequent debasements of 'modern' popular art.

This theoretical trope of the always/already lost may be understood as simply a synchronic defining feature of the popular or carnivalesque performance. On the other hand, even such moves as this may have an historically specific and changing material basis. W.D.'s discussion of stage clowning may be particularly aimed at and derived from observation of artistic pretension, class self-definition, and the commercial transactions between pantomimists and audience particular to his generation. This was the moment of the fullest flowering of the spectacular Opening and Transformation Scenes, and the press of 1859 and 1860 carries extravagant puffs for such spectacles and also the irrepressible misgivings that were soon to render them *de trop*. The *Morning Chronicle* (5 January 1859), for example, is somewhat incoherently rapturous about the 'massive spectacle of gorgeousness' of 'The Temple of the Sun, and Fairies' Harvest Home' at the Marylebone Theatre (not a large house) where

> [f]irst of all there is a dead gold scene, with perforated borders like lace-work, and pendant gold balls; then a sink in the scene discovers two fountains, covered with jewels and dead gold; and on both fountains arise two fairies, holding in their hands sheaves of corn and reaping hooks. Behind the fountains arise, for the third change, a female figure in gold lamé, holding sheaves of corn and oak. The scene at this period of its various changes is splendid, from the immense quantity

> of ossidu about it. The final rise is caused by another rise, displaying five sea-horses, really beautifully painted, and a lady in a car, driving these steeds of Neptune.

The little theatre was striving to keep up with the splendours executed at Drury Lane by the famous scenic artist Beverley, which are described with gasping enthusiasm by the *Era* in its main 1858/9 pantomime review of 2 January, 1859:

> What ... can be more refined and elegant in its conception than the 'Abode of the Arcadian Fairies?' – an ideal study of unsurpassable beauty – the trees of Porcelain crested with golden branches, which constitute the foreground, opening in the centre and unfolding a vista through which a torrent of living water is seen sporting and tumbling over a precipitous incline of rockwork. The effect of this is most charming, and when the stage is filled, as it coincidently is, with a myriad of shepherdesses, costumed after Watteau, the beauty of the picture, so happy in its composition, so dainty in its development, and so exquisitely glowing in its execution, completely ravishes the spectator, and transports him into regions so conceived, and so peopled, only in his dreams! The transformation scene, 'The Retreat of the Wood Nymphs,' is one of those elaborate mechanical combinations of which Mr Beverley ... is the sole creator. It realises itself through *media* that gradually become less and less dense, until a gigantic fern is discovered, which, in its turn, opens silently and imperceptibly, and reveals, as the scene grows, and the light becomes more effulgent, nymphs without number reposing upon the interlacing branches of a silver oak in the background, in attitudes the most graceful, and with the most perfect concealment of the means that are resorted to to produce a *denouement* at once delicate, ethereal, and lustrously, bewilderingly, magnificent.

W.D., however, in the same paper six weeks later, describes such scenes as 'like nothing under heaven, in earth, sea or air; they represent nothing, are composed of nothing, and mean nothing' except 'the best and most convenient way in which the greatest number of ladies could be suspended, enveloped, and developed by uprising and revolving machinery, whilst being comfortably surrounded by blazing gas-lights and coloured fires'. He looks forward to the time when the managers will be arrested for cruelty to dumb animals, since 'nothing less than painful torture can it be for delicate young females to be suspended in the most painful and cramping positions for so long a time, to say nothing of the great heat of the numerous surrounding gas-jets and the suffocating influence of the sulphurous red and white fires'.

The barely-disguised subject of discussion here, of course, is the sexual, titillating quality of such displays of very young women in exotic undress; but the impossibility of debating that explicitly, especially in the context of an entertainment which advertised itself more loudly each year as chiefly designed for the entertainment of children, makes the press response particularly tortuous. The theatres advertise the 'gorgeousness' of their scenes, and at the same time make favourable publicity for themselves by sometimes admitting parties of schoolchildren free.[8] The theatrical press enthusiastically reinforced this tactic, excusing the pantomimes what they chose to regard as the tedium of their gloatingly protracted transformation scenes for the joy of '[t]he silvery laughter of thousands of little gleeful voices, accompanied by the clapping of tiny hands'.[9] This manoeuvre was largely successful in disabling adverse responses, such as that of The Very Reverend Francis Close, Dean of Carlisle, who made a speech in December 1859 at a meeting of the Total Abstinence Society in which he denounced the reports of the Boxing Day pantomimes in the daily press as 'calculated to strike the heart of the Christian, and *send us all on our knees to pray to God to avert the judgement we deserved*'. His language, inevitably both lurid and inexplicit, laid him open to scorn from the more sophisticated Liberal press. The *Examiner* reports this speech[10] as a sample of staggering hypocrisy, in the face of such other reasons for repentance as the slave trade, the Cawnpore massacre and 'the great social evil' (which it does not even need to specify to its readers). The paper mocks the prejudice and envy manifested by Dr Close about the 'special temptations and wickednesses' of the theatres, which are beginning to put on two performances a day to meet demand while the church pews are empty.

Reading the pantomime as part of the mid-Victorian structure of feeling therefore, reveals something of the complexity of the situation of performance at that moment. The supporters of the stage, faced with attacks from religious fundamentalism more extreme than any that had surfaced since 1640, sought to continue the valorisation of Grimaldi's clowning as somehow linked to the innocence of children and of simple folk. But that was not easy to do, given the possibility of offending one section or another of fluctuating (and often unexpressed) public opinion. The meaning of pantomime had moved into an ambiguous place. Its cultural worth could potentially be renegotiated in terms of nostalgia, and a new understanding could replace charges of vulgarity with the (carefully-contained) indulgence of puerility. But a well-spring of uneasiness remained, and indeed had been replenished. The struggle to formulate a widely acceptable meaning for the form can be traced in the way it is treated by the critical brokers in *Punch* in the years round 1860.

The comic weekly, founded thirty years before, was by this date heading away from its initial radicalism towards the middle-brow, comfortable scepticism crossed with bursts of self-righteous sentiment with which it fed the self-satisfaction of generations of middle-class Victorians. It had recently resumed dramatic criticism, and chose to bye-line its comments on the London theatres as by 'One who pays' – repudiating the opprobrium which had become attached to the puff, and to theatre critics as the creatures of the managements who admitted them free. The readers and therefore the reviewers of *Punch* were pleased to regard themselves as the paying public, whose pockets supported the pretensions of the artistic community. An article on the season's pantomimes by 'One who pays', and sometimes an additional unattributed item making some particular point, appeared around 1860 in post-Christmas issues: for example on 7 and 21 February 1857; 6 February 1858; 8 January and 5 February 1859; 3 March 1860; 12 January and 16 March 1861; 18 January, 1 and 15 February 1862.

Until the late 1850s pantomime is still a given for *Punch*, a known cultural token not interrogated itself, but used journalistically to make a point about something else. The first of the two pieces in 1857 is about theatrical puffery: an easy joke to make, given the kind of exaggerated writing exemplified above. The columnist suggests that the seven theatres which all claim that their pantomime is 'the best' are simply conspiring to get the judicious critic to attend every one of them in order to discover which is telling the truth. The second piece concerns an incident in the news and attacks the Board of Guardians of the workhouse at Bath for refusing to allow the pauper children to accept a kind theatre manager's invitation to a free matinee performance, on the grounds that all theatre is immoral. *Punch* offers a furiously sarcastic account of this, and of the guardian's motives, suggesting that one of them had seen a play, *The Serious Family*, which attacks hypocrisy and cant, and had felt himself lampooned, so the poor little things under his charge were denied an indulgence which even the royal family felt was harmless to its children.

From this point onwards, however, the discourse becomes more complicated, and focuses upon contradictions in the form itself, as signalled by the headline of 6 February 1858: 'The Decline of the Pantomime, a lament by a critic who has worn his hair powdered' (i.e., by an old man). His preference for the experience of pantomime 'when George the Third was king' is not mocked: the column rather suggests that he is right to prefer the pantomimes of his youth, even though he smiles at his own childishness in being more interested in those days by clown's antics with the red-hot poker than he was by the plot. By 8 January 1859, however, the full-page article describing 'the conventionalities of the stage: melodrama, farce and pantomime' is presented much more ironically, and the voice assumed by the writer suggests that the topics treated are far distant from the modern tastes of the reader. The tone at the outset, when describing 'the Melodrama', could be taken to be that of a modern sensibility that finds the extremes of moral simplicity in such old plays comically naïve. When the article moves on to farces, the tone becomes apparently celebratory and nostalgic about the simple life represented by such plots and their easy resolutions. When he reaches pantomime, however, the speaker is mockingly set up as becoming clearly deranged: he avows his total and sincere belief in the reality of the world of clown and pantaloon, and breaks off in the midst of its incoherent description, as if lapsing into senile babbling, or as if the old man's script had simply got lost, or become too silly for the printer, and therefore the reader, to bother with the final paragraph.

At this point in 1859 the attitudes of *Punch* writers to pantomime waver and seem to cast about for a secure line of interpretation. The next piece, only four weeks later, goes back to the nostalgic appeal of Grimaldi as the touchstone of unquestioned comic values, but then backs away from what it at first seems to endorse. At the outset 'Give me my red hot poker! (being a plea for Cheap Pantomimes)', on 5 February 1859, demands the restoration of the ancient pantomime, what it calls 'the massive gold of GRIMALDI'S humour and art, instead of all this trash and tinsel' – pantomimes with 'the classic simplicity of the olden time' 'which shall cost the management nothing but invention in the comic scenes, and humour as well as agility and posture-making, in Clown and Pantaloon; in which thefts and slaps shall be duly insisted upon, and the butter-slide, and hot poker, restored to their proper place and significance.' The writer represents these things as what he remembers from his own childhood; but the reader soon detects some difficulty in his advocacy of their re-introduction.

He begins by saying he is not ashamed to avow a lingering love of the old pantomime – but as he moves into telling us what that is, the language begins to betray unease. In that world 'the policeman is invariably bonnetted with impunity' – not just by Clown, but 'by the mob'; 'a grotesque and grinning buffoon ... is allowed ... to commit every crime that can disgrace humanity' and Columbine and Harlequin enchanted the critic's childish senses by running away together in a 'loveflight' to – and here is the jarring note – 'common lodging-houses', where they meet washerwomen 'in a permanent state of intoxication, from gigantic bottles of gin.' Here he begins a new paragraph with the phrase 'It shakes my faith in the moral government of this Pantomimic universe' if violence towards old Pantaloon does not happen, adding 'I feel it to be perfectly in accordance with the ethics of this other and better world, that hoary but vicious imbecility should receive all the kicks, while gay but unprincipled mother-wit pockets all the halfpence. I believe in the butter-slide; I reverence the "spill and pelt"; I look upon policemen as an institution to be grossly misinformed, scoffed at, and smitten.' This creed is very obviously not only ironic, but also rather troubled. The old practices of pantomime are relished; but, in reality, they are felt to be no longer suitable perhaps, especially for children. The rational adult must actually repudiate an entertainment which is violent, immoral and altogether vulgar.

On 3 March 1860, the month after the cartoon of the children, the journal prints another full-page eulogy of the 'good old comic clown' after Grimaldi, which like the drawing recognises nostalgia as a widespread response to pantomime and makes that its central joke. The piece is absurd in its earnestness, and culminates in preposterous nationalism about rejecting foppish French Pierrots in favour of 'the fine old JOE GRIMALDI style'. It is signed as by 'An Englishman, and one of the old sort', who is thereby distanced as a correspondent, and by implication not a *Punch* professional. This resolution of the conflict does not last, however: the journal cannot now leave nostalgia behind and judge the modern Christmas shows without reference to the past.

In the next year a new resolution is sought, and one which can distance the readers and writers of *Punch* from a form becoming unacceptable in all its manifestations, past and present. On 12 January 1861 an effective, indeed trenchant, set of comic verses appears, entitled 'A Lament of the Season, by an old Boy':

I used to think, when I was young,
That fairy-land was fair;
But now I know 'tis tinsel all,
With red-fire fumes for air;

A land of traps and pitfalls,
A land of sloats and stays,
And scruto-work and profiling,
And shivering coryphées.

Where are the joys of Pantomime,
I knew in days of yore?
The poker hot – the butter-slide –
The Clown laid at the door,
That shopmen bland and affable
Might o'er him tumble down –
The murdered babe, whose body
Was sat on by the Clown?

The contrasted pantomimes, past and present, are both inhumane, or worse. In March 'One who pays' files a report on his theatregoing, and finds a way to set pantomime apart not simply as belonging to the past, but as belonging primarily to another class altogether, which can best be approached in his middle-class professional capacity with a kind of judicious detachment, as a way of demonstrating his own expertise and knowledgeableness about theatrical practice. He has been to the lower-class theatre in the City Road, and condescends to 'own that I laughed more at the Royal Grecian pantomime than I have done' at many places further West. He ends his notice by congratulating Mr Conquest, the manager, and his audiences, for knowing that Clown is essentially a thief and a fool, who burns himself on the poker and gets caught because he leaves the fish he has stolen hanging out of his pocket. Thus put in its place, as an amusement for the unsophisticated which can be trusted not to corrupt them, if properly managed, pantomime is acceptable. We do not, however, need to attend it ourselves.

But, of course, we still do; so an excuse is required, and one which covers the potentially embarrassing fact that many middle-class theatregoers – especially, perhaps, male ones – pack the theatres that offer the transformation scenes, and apparently enjoy the spectacle of 'shivering coryphées'. In 1862 the *Punch* dramatic correspondent handles this by using a convenient trope that sets down the pantomime as a show for children of two sizes: real children, who have a simple and inexpensive relish for the red hot poker; and 'grown up children', who are so vicious in their tastes that to please them 'the scenery each Christmas is more prominent and splendid, and every year the stage effects are made more highly spiced'. He then tacitly dismisses this 'spicy' entertainment as not worthy of any further critical attention, and focuses on describing the Christmas shows for young people. There is still a shiver of ambiguity, however, when he gets round to recommending the Crystal Palace for Christmas, where the father of a family can spend conspicuous and inappropriate amounts of money on toys and sweets, and take his children to see music hall turns such as 'the Great Mackney' in black-face, warbling songs that would fall flat without it, and 'the Perfect Cure' persuading his audience to applaud 'the worst of music' by bouncing up and down while he sings it. Patronising these hugely popular entertainers is in itself a relaxing of strict class and taste boundaries, since it is assumed that neither the middle-class father nor, of course, his children would normally frequent music halls. The indulgent papa at the Crystal Palace may conclude the evening by letting the over-excited young ones see Blondin in a 'pantomimic drama' as a monkey, tempting them to go home and walk up the banisters, jump across the drawing room from chair to chair, or play at football with the baby. Thus the critic recommends a series of entertainments that he simultaneously critiques and deprecates. The article implies that exposure to popular entertainment like pantomime is a traditional and necessary experience for the young. Its moral and cultural dangers must be resolutely managed. *Punch* assists with that process in such articles as this by adopting an ironic tone that distances the consumer, the reader of the magazine, from enjoyment of conspicuous consumption and dubious stage performances while giving him permission to indulge in them under the guise of initiating his young ones. It is a tricky procedure, and one which shows how complex a balancing act the pantomime is performing.

The picture of the experienced young fellow shows, perhaps, how difficult such ambiguities are to dissipate or assimilate entirely. The boy is assuming the role of elder, the old theatregoer who has seen the best days of the pantomime, while the little girl is possibly the disappointed child of whom the articles speak: she has been to the pantomime and been bored by the slow transformation scenes and disappointed by the absence of violent and thieving clowns with red-hot-pokers. The reader is invited to laugh at the little fellow for setting himself up as an expert on something he can't even pronounce; but more than that is suggested by his depiction in the pose of the lecturing adult, standing over his sister with a wagging finger raised – but with his other hand plunged in the pocket not of a pair of trousers, but a skirt. His tartan socks suggest this might be a kilt, and in any case young boys up to the ages of five or six were still sometimes not breeched at this period; but the joke about his assumption of adult knowingness nevertheless sits uneasily with his lack of masculinity in dress. His long hair falls effeminately over his face; the image is inescapably gender-ambiguous. Meanwhile Clara, undoubtedly a girl, behaves in a very unladylike manner. Slumped in a chair clearly not meant for her to occupy, hands negligently resting upon the arms, she dangles her legs in full view, clad in nether garments not unlike those of the principal boy. Showing her garters on her thighs and crossing her ankles like a lad, she confronts male admonition unabashed, unimpressed, with eyes levelled at his waistcoat and a discontented scowl, or perhaps an ironic smirk. The body-language of the quasi-cross-dressed children speaks volumes about the part that pantomime and its consumption played in Victorian culture, on a level more hidden, less conscious, but arguably more subversive than breezy hypocrisy about half-dressed chorus girls and populist mock-despair about modern music.

Notes

1. Derek Salberg, *Once Upon a Pantomime* (Luton: Cortney Publications, 1981), pp. 76–83, p. 83.
2. Millie Taylor, *British Pantomime Performance* (Bristol: Intellect Books, 2007), pp. 22, 21.
3. Morag Schiach, *Discourse on Popular Culture: Class, Gender and History in Cultural Analysis 1730 to the Present* (Cambridge: Polity Press; and Stanford: Stanford University Press, 1989).
4. Maurice Willson Disher, *Clowns and Pantomimes* (London: Constable, 1925), p. 248.
5. For a fuller account see John O'Brien, *Harlequin Britain: Pantomime and Entertainment 1690–1760* (Baltimore: Johns Hopkins University Press, 2004).
6. David Mayer III, *Harlequin in His Element: The English Pantomime, 1806–1836* (Harvard: Harvard University Press, 1969).
7. For a more extensive analysis of the clown, see Jacky Bratton and Ann Featherstone, *The Victorian Clown* (Cambridge: Cambridge University Press, 2006) and for workshop experiments with Victorian clowning visit: www.rhul.ac.uk/Drama/Research/video/clownvideo.html
8. *Reynolds's News*, 29 January 1860, for example, reported that the Drury Lane Pantomime had hosted 126 children from the Licensed Victualler's School, and that those of the Duke of York's Military School would be coming next week.
9. *Era*, 9 January 1860.
10. *Examiner*, 8 January 1859. The quotation given appears in this report.

6

'Only an Undisciplined [Nation] would have done it': Drury Lane Pantomime in the Late Nineteenth Century[1]

Jim Davis

In the 1880s and 1890s the most spectacular pantomimes on the English stage could be found at Drury Lane, under the management of Sir Augustus Harris, whose policy of presenting a spectacular melodrama every autumn and a spectacular pantomime every Christmas was proving unfailingly successful. Not only were his pantomimes renowned for their spectacle and processions, but they also employed the most popular of contemporary music-hall stars in leading roles – an innovation for a West End theatre of Drury Lane's status, although the presence of music-hall performers from the late 1860s in pantomimes at other theatres, such as the Britannia in East London, was long established.[2]

The effectiveness of using comedians, singers, speciality acts and even minstrel acts from the halls was thus established by the time Harris moved in on the scene. What was new was the fact that the allegedly middle-class families who visited the Drury Lane pantomimes would not normally have frequented the music halls: to see such stars as Marie Lloyd, Herbert Campbell, Dan Leno, Little Tich or Ada Blanche was therefore to get a taste of usually forbidden fruit. In employing such performers Harris was drawing on the cream of the music-hall profession. Leno, with his slight form, wan face, and eccentric personality, was perfectly teamed with Herbert Campbell, who, at over six feet in height and over twenty stone in weight, was his complete physical opposite. Marie Lloyd, well known for her *risqué* songs, and Little Tich, famed for his comic dancing in grotesquely elongated boots, were equally popular.

Under Augustus Harris, the pantomime text became more and more subsidiary. E. L. Blanchard, author of many Drury Lane pantomimes up until the 1880s, complained that his texts, over which he took enormous pains, completely disappeared to make way for scenic effects, music-hall songs, and the comedians' own ad-libitum material.[3] Slang, innuendo, and inebriation became features of the new pantomime: when Marie Loftus played Little Bo Peep, it wasn't the sheep that strayed but Bo Peep herself. W. Davenport Adams had complained as early as 1882 that the music-hall performers, engaging in pantomime,

> bring with them, so to speak, an atmosphere which it is sad to see imported into the theatre. They bring with them not only their songs, which, when offensive in their wording, are sometimes made doubly dangerous by their truthfulness – not only their dances, which are usually vulgar, when they are not inane, but their style and manners and 'gags', which are generally the most deplorable of all. The objection to music hall artists on the stage is ... that they have the effect of familiarizing audiences, and children especially, with a style and a kind of singing, dancing and 'business' which, however it may be relished by a certain class of the population, ought steadily to be confined to its original habitat.[4]

The snobbery and social prejudice of this statement are fairly explicit and raise further questions about exactly what transactions were occurring, between audience and performance, in the course of the Drury Lane pantomimes.

War and militarism: *Jack and the Beanstalk*

Like the Harlequinade at the beginning of the nineteenth century (not to mention most subsequent pantomime entertainment), the Drury Lane pantomimes also included topical material. Dan Leno and Herbert Campbell usually sang a topical duet each year in which much of the material was added on the spur of the moment, despite the fact that the 'official' pantomime script still had to be submitted to the Lord Chamberlain's Office for censorship each year. Topicality, however, was not merely the province of the comedians. In *Sinbad the Sailor* (1882–83, E. L. Blanchard) a procession of English Kings and Queens was greeted by 'the conquering heroes from Egypt' – boys and girls dressed to represent the British army which had just been victorious in the Battle of Tel-el-Kebir. The militarism of the 1890s was represented in the opening scene of *Dick Whittington* (1894–95, Augustus Harris, Cecil Raleigh and Henry Hamilton), which displayed a series of military manoeuvres performed by a regiment of feline soldiers and a variation on a popular theme in the song of 'Tommy Catkins' (Figure 6.1). But perhaps the most notorious example of militarism in Drury Lane pantomime

Figure 6.1 Feline Soldiers from *Dick Whittington,* Drury Lane, 1894–5 (collection of Jim Davis)

occurred in *Jack and the Beanstalk* (1899–1900, Arthur Sturgess and Arthur Collins), when the British were at war with the Boers.

The Boer War had commenced on 12 October 1899, nominally over the rights of British subjects in the Transvaal to be enfranchised as South African citizens. President Kruger, the Boer leader, resisted the British demands, fearing acquiescence might lead to eventual British domination of the Transvaal. The British in turn recognized the strategic importance of South Africa to their Empire and, wishing to show they were an adversary not to be trifled with, declared war on the Boers. During the first three months of the war they were less successful than they had expected, suffering a series of defeats which culminated in the notorious 'Black Week' of the 10th–17th December. Later, when Boer commandoes waged a sort of guerrilla warfare against the British troops, the British indiscriminately employed a scorched earth policy (burning Boer crops and homes and destroying Boer stock) and invented the first concentration camps, in which to intern displaced Boer women and children. Nowadays there is general agreement that the Boer war was unnecessary, dishonourable, incompetently managed by the British and an unpleasant manifestation of imperial aggression by a nation slowly going into decline. At the time, when the extension of British power abroad was perceived as a god-given right and those who opposed it must be either traitors or savages, there was enormous enthusiasm for the war, reflected in the many Boer war melodramas and patriotic songs that emerged. After 'Black Week', Lord Roberts, a hero of the Indian Mutiny, was appointed commander in chief of the British forces in South Africa on 18 December 1899, and Lord Kitchener, the hero of the more recent Sudan campaign, became his chief of staff. Patriotic frenzy knew no bounds and, not surprisingly, the annual Drury Lane pantomime also contributed to the deluge of patriotic fervour and anti-Boer propaganda.

Jack and the Beanstalk was first performed on 26 December 1899. The diminutive Dan Leno played Dame Trot, Jack's mother, and Herbert Campbell her younger son. Jack was to be played by Nellie Stewart, a star of the Australian musical stage, but on the first night she withdrew, allegedly suffering from laryngitis, and her role was undertaken by Mollie Lovell and then by Violet Cameron for the rest of the pantomime's run. The book of the pantomime does not read well today: the story line seems convoluted and unclear and directions for the set spectacle scenes, a major feature of the entertainment, are not provided. *The Times* (27 December 1899), however, described some of these set pieces, including a market scene with an impressive vegetable ballet in which 'children [dressed] as leeks dance[ed] to the music of Men of Harlech, potatoes and carrots to Irish tunes, and so forth, *ad infinitum*, ending with a pretty dance of girls with headpieces of broccoli'. The first half of the pantomime concluded with another set-piece, the *Land of Harmony*, a homage to Wagner, with knights on swans à la *Lohengrin*. The finale to the entire show consisted of a tableau entitled *The End of the Century* in which was 'represented, in beautiful allegorical fashion, all the wonders, discoveries and inventions of this, the best of all centuries'.[5]

The printed script contains only a couple of references to the Boer War and to conflict in general. In the first scene Jack's forthcoming fight with the giant is viewed as a patriotic duty in which victory is assured and concludes with a patriotic song beginning with the lines:

> Soldiers of the Queen
> Fighting for country heart and soul …

In the second half of the show the Giant tells Jack's sweetheart, the Princess, whom he has imprisoned in his dining-room, that they will have a dance, to which the weeping Princess replies:

Princess: Your step mayn't suit me
Giant: Don't you cry and snuffle.

Princess: What are you good at?
Giant: Why the double shuffle
Princess: Your shuffling English people can't impugn
Giant: They made me dance, and to a pretty tune.
Princess: They owe you one – they've kept their kith and kin
All in your pocket.
Giant: That was a Lady's myth.

The point of this doggerel exchange is that the Giant Blunderbore had been crafted to resemble the South African leader, President Kruger, and turned into the more topical *Blunderboer*. The Princess is referring to what the British saw as his prevarications and to his boast that he could keep the British army in his pocket. Ladysmith had been the site of a decisive Boer War battle. There may have been further references to the war in a scene during which the comic characters 'discuss various matters of urgent public importance'; however, this is all the book of the play tells us and since this was the version also submitted for the censor's approval, whatever else was said was said ad libitum. But the core scene in this pantomime was 'The Fall of the Giant', in which, after crashing down from the beanstalk, the fifty-foot body of giant Blunderboer lay sprawled across the Drury Lane stage. The *Sketch* (27 December 1899) described how, as the lights came up:

> Not only does the hero Jack stand upon the giant's knee and warble an appropriately defiant patriotic ditty [entitled *Pretoria*] – while War-Correspondents, seated on his boots, prepare their copy – but from a pocket of the Giant Boer there issues forth a mighty army of British and colonial troops, each one realistically uniformed and wonderfully trained.

All the troops were played by children, defiantly disproving Kruger's boast that he could keep the little British army in his pocket. *The Times* (27 December 1899) described how:

> Miniature British troops, some in the scarlet of infantry, some in khaki [a new uniform first introduced the previous year], some gunners, some blue-jackets with a machine-gun, some mounted infants in the strict sense of the word [on miniature Shetland ponies], and some Australian lancers throng the stage and walk over the body of the giant.

The Times also referred to the enthusiasm that broke out when the 'childish soldiers raised their helmets on their rifles' and everyone burst into *Rule Britannia*. The same newspaper concluded its review of the pantomime by asserting that 'Drury Lane promises to maintain its traditional hold upon the affections of the children'. For, we must remember that these pantomimes were described as children's entertainments and targeted child spectators.

In presenting such scenes the Drury Lane pantomime appeared to celebrate Britain's imperial strength and to support or encourage patriotism within its audiences, albeit through feminized heroes and diminutive regiments of children. The very splendour of their presentation marked out these pantomimes as a celebration in its own right of British supremacy, although in 1900 this tendency was criticized by *The Star*:

> The Drury Lane pantomime, that national institution, is a symbol of our Empire. It is the biggest thing of the kind in the world, it is prodigal of money, of invention, of splendour, of men and women; but it is without the sense of beauty or the restraining influence of taste. It is impossible to sit in the theatre for five hours without being filled with weary admiration. Only a great nation could have done such a thing; only an undisciplined one would have done it.[6]

The *Star's* description of the Drury Lane pantomime as a national institution raises certain problems, not least of which is the identification of the Drury Lane audiences and the systems of belief they represented. The status of Drury Lane – historically, economically, as surrogate national theatre and as a centre of entertainment at the heart of a great imperial power – also deserves consideration. We also need to ask what the engagement of certain music hall performers meant: not just what they did on the Drury Lane stage, but what (through their prior reputations) they signified.

Wealth, poverty, morality, gender

Drury Lane cannot be defined as a neighborhood theatre by the end of the nineteenth century, nor can the audience be specifically located, since improved transport facilities had inevitably opened up the theatre's accessibility to a wider network of patrons. The scripts of the pantomimes themselves, although they survive, were heavily cut and often distorted in performance by the introduction of ad-libbed material.

Thus discussion of their circumstances of performance and assessment of any ideological transactions mediated through them are likely to be speculative. None the less, issues of class, imperialism, and patriotism, and the representation of race and gender are implicit (at least for the modern investigator) in texts, illustrations and contemporary accounts of many of these pantomimes.

The pantomimes discussed below were performed during the 1890s and bridge the transition of management after Augustus Harris's death in 1896 to Arthur Collins. Their tone was distinctly middle class, in so far as they tended to mock both the idle rich and the irresponsible poor. Thus, in *Babes in the Wood* (1897, Arthur Sturgess and Arthur Collins) Prince Paragon turns out to be a brainless swell with the catch-phrase 'Don't cher know?', while the second part of the pantomime opens with a chorus of Piccadilly Johnnies singing, 'What's the use of being bwainy?' In *Little Bo Peep, Little Red Riding Hood, and Hop o' My Thumb* (1892–3, Augustus Harris and Wilton Jones), on the other hand, the poor are presented as feckless, incompetent, and neglectful. Daddy Thumb (Dan Leno), for ever trying to be rid of his children, exclaims:

> Once for all, are we working men not to have as many children as we like, and six times as many as we can afford to keep? Aren't we to have the privilege of leaving them in woods and dropping 'em down 'aireys' without any bloated government interfering?

He is later rebuked by his wife, Goody (Herbert Campbell):

> If you hadn't spent so much in the public house we shouldn't be under the necessity of losing our poor children now.

Attempts at losing them prove difficult, however, and they are reminded by one of their offspring, Hop o' my Thumb (Little Tich), that child desertion means 'six months with hard labour'. Daddy later sings about his participation in the 'Midnight March':

> You've read it through in *Lloyd's*,
> The so-called Un-employeds,
> Had rare work to do the Midnight March.

The song also comments on what 'a bootless farce' it is for the unemployed to try and improve their lot. In *Cinderella* (1895–6, Augustus

Harris, Cecil Raleigh and Arthur Sturgess) the reverse appears true, as one of the songs, 'The Up-to-date Domestic', celebrates the fact that:

> Progressive legislation and the School Board and the Press,
> Now are lev'lling up the masses with remarkable success ...
> It's a very thin partition now that class from class divides.

However, since the insinuation behind the song seems to be that servants no longer know their place, there is something of a sting contained in these sentiments. Not only are the poor, the unemployed and the domestic targeted: in *Little Bo Peep* the Wolf is not only a drunkard but a socialist, proclaiming, as he staggers from The Spotted Sun, 'I'm a Socialist, and my motto is "Down with anybody who has got any money, and up with those who haven't"'.

Drinking and flirting are recurrent characteristics of these pantomimes. Little Boy Blue, the Wolf, Red Riding Hood's Grandmother, Daddy and Goody Thumb are all fond of their tipple in *Little Bo Peep*. Dandini sings a song in praise of immoderate consumption of wine in *Cinderella*. In *Dick Whittington* Idle Jack and Cook comment on the fact that everything in London is closed on Sundays except the public houses: going to the pub is the only pursuit allowed in 'moral London on a Sunday'. Drunkenness is endemic throughout the Drury Lane pantomimes, both in its attribution to certain social classes and as a source of comedy.

If one assumes that Drury Lane audiences were for the most part exposed to behaviour that they would not countenance in everyday life, a voyeuristic tendency emerges. This is perhaps apparent in the many flirting scenes that occur, most notoriously in *Little Bo Peep*, concerning which the *Illustrated Sporting and Dramatic News* (14 January 1893) complained of the 'occasionally risky passages' it contained. The appearance of a gallery of famous nursery rhyme characters takes us into a very different world from that represented in their sources:

> *Enter* TOM TUCKER
> TOM T. Here just come along Jack Horner.
> *Enter* JACK HORNER.
> I saw you kiss Miss Muffitt round the corner.
> TOM P. [TOM THE PIPER'S SON]: Oh, here's a lark! You won't believe it's true
> Red Riding Hood is *flirting* with Boy Blue!

ALL (*incredulously*). No!
Enter DICKORY DOCK.
DICK. They're squeezing – kissing!
MAR DAW. For a squeeze I'm busting.
MISS MUFFITT. I call such shameless goings on disgusting.

A little later on all the female nursery rhyme characters overpower the boys and hold them in position so that they can kiss them.[7] An additional encounter, between the Prince and Bo-Peep, leads the Dame, Mary Contrary, to declare that 'this mania for kissing seems universal. The County Council ought to put it down'. The *Athenaeum* (31 December 1892) might have agreed, considering that the engagement of music-hall 'artistes' added 'spirit' to the action of the pantomime, but at the expense of 'delicacy and charm':

> What is worse, the stories in which childhood delighted are vulgarized. To see Little Red Riding Hood or Little Bo-peep presented by a young lady with the pronounced style of the music-hall, to hear her talk of nothing but kissing and hugging, and to watch Little Boy Blue tipping a knowing wink to his sweetheart for the time being, is nothing short of desecration. Oh, Sir Augustus! ... leave our children their fairy tales ... leave us the nursery and the cradle ... your little shepherdesses of the music-hall, with their vulgar fancies and style, have nothing to do with pantomime nor with childhood![8]

Yet just two years later a scene set near Highgate in *Dick Whittington* actually required lads and lasses to frolic in the hay. Such scenes as this and those in *Little Bo Peep* were obviously bringing the music-hall tone into the Drury Lane pantomimes, although the fact that all the male and female roles in these instances (other than Dame Mary Contrary) were played by women may have imbued them with a certain innocence in the eyes of contemporaries (if not in our own.)

The Drury Lane pantomimes were relatively conservative in stance, although the approach varied from issue to issue. Thus the London County Council's (LCC's) strictures on morality and drunkenness and Mrs Ormiston Chant's moralistic assault on the music halls in 1894 (aspects of which included the alleged presence of prostitutes in the promenade of the Empire Theatre of Varieties, Leicester Square, and the supposedly indecent costumes worn by its female dancers) were treated fairly satirically.[9] Fitzwarren in *Dick Whittington* protests at the way in which the LCC has put an end to the good old days of wining and dining at the Guildhall:

> It's stopped our institution, seized our funds,
> And banquets us on tea and sally-lunds,
> With magic lantern and a glass of milk,
> The Lord May'r entertains; our robes of silk
> Have all been changed to something that will wash,
> The loving cup is filled with lemon squash.

'The Song of the Municipal Theatre' sung by Ladies of the Ballet in the same pantomime also protests at the current concern with moral rectitude:

> At the Municipal Theatre we're engaged to dance,
> Because we've never seen the naughty things they do in France,
> We cannot turn a Catherine-wheel, or kick, or split, or prance …
>
> At the Municipal Theatre nobody can see
> A single thing to shock a guiltless infant under three.
> Silk tights we sternly stop with shan't spelt C.H.A.N.T.
> There's nothing stronger sold to drink than coffee or Bohea,
> And all the jokes and equivokes are proper as can be!

The song concludes with a reminder that this sort of thing was tried once before, in Cromwell's time, but the Restoration of Charles II soon led to the restoration of mirth and revelry. A. E. Wilson describes how this scene was staged:

> [A] group of girls from the Municipal Theatre in Quakerish gowns and poke bonnets of somber colour danced a mock measure with downcast eyes until, carried away by their feelings, they broke into a wild can-can to Offenbachian strains and finally performed somersaults in the manner of the Bal Bullier.[10]

The LCC's enquiry into immorality in the music halls is also satirized in the ensuing dialogue between Dick and Idle Jack towards the pantomime's end:

> DICK: But – dear me – what is a Music Hall?
> JACK: Well, my Lord, it's a kind of hall – a hall –
> DICK: Where people go to hear music?

JACK: Oh dear, no! Where people go to walk about, and see as little of the show as possible.
DICK: Then a Music Hall is not a place of entertainment?
JACK: It has been so described, my Lord.

A reference to the music hall issue also creeps into the following year's pantomime, *Cinderella*, when the Prince (in disguise) tries to arrange an outing with Cinderella:

PRINCE: Where shall we go? Say, to a Music Hall?
CIND: Oh! Dear me, no; that wouldn't do at all!
For months I haven't dared such sights to see.
PRINCE: They're all right now. They're as they used to be.

The Drury Lane pantomimes are inevitably on the side of the halls – not surprisingly, considering how many of each cast was recruited from this source.

Less sympathetically treated were the New Woman and the New Drama. In *Dick Whittington* Herbert Campbell as Alice the Cook appeared dressed in deliberate parody of fashions favoured by the New Woman. (Figure 6.2). Idle Jack subsequently offers to marry her, and tells her she shall have:

Everything a New Woman can desire – you can grow yellow asters and green carnations in the back garden ... You can join the queer club and I can nurse the baby. Your father shall live with us. You shall have a happy and united home, and a divided skirt.

In *Babes in the Wood* the appearance of a character named Gertie Girton proved the new woman was not extinct and in *Cinderella* the Demon proclaimed:

I gloat on every game I play to watch the way it rubs
I introduced New Women and suggested Ladies' Clubs.
I brought out on the bicycle for all folk to admire,
The *fin de siècle* lady with the masculine attire.
I patronize the Drama too, the healthy play to blast,
And hinted to the Dramatists the Woman with a Past,
It made the people wonder if they really were awake,
And the critics writhed and wriggled, but I laughed to see 'em quake.

Figure 6.2 Herbert Campbell as the 'New Woman', *Dick Whittington,* Drury Lane 1894–6 (collection of Jim Davis)

Cinderella also claims that problem plays send her to sleep. Indeed, Drury Lane's theatrical fare is the most highly commended in this pantomime, with references not only to the success of its annual Christmas entertainments, but also to one of its most successful melodramas, *Cheer Boys Cheer*. In the opening scenes of *Babes in the Wood* the claims of Drury Lane pantomimes are also measured against the rival attractions of Ibsen, Shaw, and French farces.

Representing race: *Robinson Crusoe*

In 1893 *Robinson Crusoe* (1893–94, Augustus Harris and Harry Nicholls) was the source of the annual Drury Lane pantomime. If, for modern readers, the novel itself raises questions about economic individualism and colonial relationships, this Drury Lane adaptation unsubtly maintained a sense of imperial supremacy and the re-enforcement of racial stereotypes. Not only did Little Tich, a midget, play Man

Friday, but additional black-face characters included the King of the Cannibal Islands, together with his Queen and Prime Minister (played by LeClerq, Brown and Newland, members of a well-known minstrel act), and Friday's father, Thursday. The tone of the portrayal of racial difference in this pantomime was complemented by an advertisement, which is printed alongside the published text, for S. and H. Harris's Ebonite Blacking for Boots and Shoes. The advertisement for the blacking includes the company's trademark, a portrait of a black woman, beneath which is inscribed 'Black but Beautiful'. Illustrations within the text depict the King, Queen and Prime Minister, as well as John D'Auban blacked up for a dancing role, as caricatured negroes. Man Friday, however, although blacked up, retains recognizably Caucasian features. (Figures 6.3a,b and 6.4).

The depiction of these characters obviously draws on the minstrel traditions of the times, while also accentuating certain racial stereotypes such as cannibalism and polygamy. On his first appearance Man Friday is confronted with the Queen of the Cannibal Islands, determined to match her husband's polygamy by adding Friday to her consorts. Apprehended by the King, Friday is in danger of being roasted at the spit

Figure 6.3a/b Little Tich as Man Friday and John D'Auban as Noblulu, *Robinson Crusoe*, Drury Lane 1893–4 (collection of Jim Davis)

Figure 6.4 Brown, Leclerq and Newland as the Queen, King and Prime Minister of the Cannibal Islands, *Robinson Crusoe*, Drury Lane 1893–4 (collection of Jim Davis)

to provide the King's dinner, but is rescued by Robinson Crusoe after (somewhat ironically) pleading:

> Protect me! And I'll be your humble slave!
> I know you will! For you look good and brave!
> And if I live, I'll do as white men do.
> I'll sometimes wash! And put on clothes like you!
> I'll wear high collars, boots, and smoke cigars!
> And drink, and flirt with ladies over bars!
> I'll turn day into night, and shout and sing,
> The latest comic songs, and front bells ring,

And wrench off knockers and assault the p'lce
And get bound over too, to keep the peace!
Save me! All this I'll do! You'll be surprised,
How soon the Savage will be civilized!

Evidently, the black-face tradition in which Friday is placed, while endorsing racial stereotyping, also permits social satire. The naming of Friday, his economic relationship with Crusoe, and requirements regarding his attire, which are all transacted in the ensuing scene, partake further of this tradition:

CRUSOE. Now, what's your name poor Blacky?
FRIDAY. 'Kill-ee-gowollop-um-skully-go-cracky!'
CRUSOE. Yes, that saved trouble. It's a pretty name;
I'd like it shorter, if it's all the same!
Let's see! Today to fry you your friends tried – eh?
So in that sense it might have been your *fry*-day,
And this *is* Friday – so I'll call you that!
Now as to terms. Of course it's understood
You'll have no wages for the first half year,
Or till you've learnt your work. Instead of beer
You'll have the money – if you're lucky!
FRIDAY. Yes!
CRUSOE. You must bestow more care upon your dress;
When visitors may call. (*Whispering.*) Wear trousers – sometimes.
FRIDAY. Of course I'll do whatever you request!
Golly, shan't I look funny when I'm dressed!

There follows an 'Indian Ballet', at the conclusion of which a chorus of Amazons exhort their 'brudder savages' to prepare the fire and cooking pot for the next meal (Friday's father, Thursday) for the cannibal King and Queen. Thursday appears resigned to his lot, but begs the royal family:

Then cook me carefully – at least!
Serve me up nicely – baste me well, of course;
Plenty of salt and pepper – mind the sauce!

A 'Nigger Boys' Chorus [*sic*] commences, after which Crusoe and Friday enter and rescue Thursday. Towards the pantomime's conclusion, when all the characters arrive in England, the King of the Cannibal Islands decides to abdicate and domicile himself there, only too happy to relinquish his native lands (and his Queen). In order to acclimatize the King and his retinue to English life, he is soon to be taken to 'Appy 'Ampstead:

> We'll take our dusky friends about
> Old London town to see!
> Their savage minds will be impressed
> With all they see, you bet,
> But there's one place above them all
> That we must not forget,
> That's 'Ampstead, 'Appy, 'Appy 'Ampstead.

Underlying the jokes and repartee is an implicit sense of national superiority. The notion of England as the centre and the rest of the world as peripheral and 'other', recurs frequently.

The mixture of patriotic disdain and smug amusement with which other nations (especially their languages) are depicted is implicit in further references within this pantomime to the Chinese, the Japanese and the Germans. At the pantomime's conclusion another of Harris's famous processions enhanced the sense of nationhood when a grand procession of Kings and Queens from William the Conqueror onwards took place, culminating inevitably in the appearance of Victoria, 'Queen of Great Britain and Empress of India'.

Conclusion

The Drury Lane pantomimes were predictable in the emphases they placed upon different topics. The poor, the unemployed, the socialist, and the idle swell are all pilloried, as are the LCC, progressive drama, racial difference and female emancipation. The music-hall tone, on the other hand, is exonerated, while slang, inebriation and harmless flirtation are presented unapologetically. It may be that the sheer spectacle of Drury Lane pantomime subsumed everything else, but it is just as likely that these entertainments had a wide influence in endorsing or confirming social attitudes and prejudices. Certainly, pantomime audiences changed during the Harris and Collins era. Jimmy Glover, musical director of the Drury Lane pantomimes, recalls how, in the pre-Harris days, the pantomime audiences consisted mainly of a huge eightpenny pit and a mass of sixpenny and fourpenny galleryites, who 'catcalled' the opening on Boxing Night and shouted the latest music-hall songs. He continues:

> The better educated audience of today is not the chorus-singing urchin or patron of the seventies. The top gallery disappeared under the LCC during the end of the Harris management, and the stalls on Boxing Night have developed into a kid-gloved army of *dilettante* patrons, across whose apathetic, well-dined and gloriously-gloved personalities, it is a far run to get the pit for a good honest round of applause.[11]

The *Daily Telegraph* (28 December 1897), reviewing *Babes in the Wood*, regretted the passing of the old, boisterous gallery:

> We could not help noticing last night that the gallery of Drury Lane was wonderfully subdued. It was in high spirits, it could whistle, it could give undoubted evidence of vocal ability when necessary, but, speaking generally, there was an air of decorum about it that contrasted strongly with olden times.

Not only was the gallery restrained. The Drury Lane pantomime was now drawing fashionable spectators (especially on Boxing Night) and respectable suburban families. It is also clear that, however strongly or not patriotic appeal was built into the pantomimes, other traditions existed to exhort patriotic sentiment in the audiences. Uniquely, 'Rule Britannia' always followed the National Anthem at Drury Lane pantomimes and, on Boxing Night, someone was deliberately planted by the management in the auditorium to hiss during the National Anthem, in order to arouse the entire house to patriotic frenzy.[12]

Drury Lane pantomime in the late-nineteenth century was in many ways hegemonic, coercive and indicative of the yellow-press tone which was about to dominate popular journalism. Yet it also employed music-hall singers who subverted, in every possible way, the respectable middle-class tone to which it arguably aspired. After all, it was Herbert Campbell who sang, in response to G. H. Macdermott's patriotic ballad 'I don't want to fight, but by jingo if I do', 'I don't want to fight, I'll be slaughtered if I do'. It was Marie Lloyd who had built her reputation around innuendo and double-entendre. Of course, these performers may themselves have been subverted by Drury Lane, although the semiotic message of their on-stage presence is, at the very least, ambiguous.

Did such performers succumb to the lower middle-class tone of Drury Lane pantomime or did they help to perpetuate it?

The tone of music hall itself had changed with the rise of middle-class respectability in the 1880s and 1890s.[13] The composition and beliefs of the Drury Lane audiences would have represented a wide range of perspectives by this time, so that the transactions occurring between stage and auditorium during the annual pantomimes are difficult to decipher. Adults and children, fashionable and suburban spectators, must have responded on a multiplicity of levels. And, within this context, wasn't there something rather absurd, as *The Star* implied, about a nation which celebrated its imperial achievements and supremacy most effectively through pantomime, with its tinsel, spectacle, music-hall performers, parodic representations of difference, transvestism and parochial lower middle-class tone?

Notes

1. An earlier version of this chapter originally appeared in *New Theatre Quarterly* 12.46 (May 1996), pp. 147–55, under the title 'Imperial Transgressions: the Ideology of Drury Lane Pantomime in the Late Nineteenth Century'.
2. See Jim Davis, ed., *The Britannia Diaries 1863–1875: Selections from the Diaries of Frederick C. Wilton* (London: Society for Theatre Research, 1992).
3. *The Life and Reminiscences of E. L. Blanchard*, ed. Clement Scott and Cecil Howard (London: Hutchinson & Co., 1891), Vol. 2, pp. 528–97.
4. See *The Theatre*, n.s. vol. 5 (February, 1882), p. 88.
5. *The Sketch*, 27 December 1899.
6. *The Star*, 27 December 1900, quoted in Michael R. Booth, *Victorian Spectacular Theatre* (London: Routledge, 1981), pp. 89–90.
7. This episode has been deleted in the prompt copy held in the Theatre Collection, Victoria and Albert Museum.
8. Quoted in Midge Gillies, *Marie Lloyd: The One and Only* (London: Orion, 1999), pp. 75–6.
9. See Joseph Donohue, *Fantasies of Empire: The Empire Theatre of Varieties and the Licensing Controversy of 1894* (Iowa City: University of Iowa Press, 2005).
10. A. E. Wilson, *Pantomime Pageant* (London: Stanley Paul, 1946), p. 88.
11. James Glover, *Jimmy Glover: His Book* (London: Methuen & Co., 1911), pp. 138–9. For a discussion of Boxing Day and Boxing Night audiences at Drury Lane and elsewhere in the nineteenth century see Jim Davis, 'Boxing Day' in Tracy C. Davis and Peter Holland, eds., *The Performing Century: Nineteenth Century Theatre's History* (Basingstoke: Palgrave Macmillan, 2007), pp. 13–31.
12. Glover, *Jimmy Glover His Book*, p. 139.
13. Peter Bailey, *Leisure and Class in Victorian England: Rational Recreation and the Contest for Control 1830–1885* (London: Methuen, reprinted 1987), p. 174.

7
Dan Leno: Dame of Drury Lane

Caroline Radcliffe

Pantomime invokes confusion and paradox, emphasising the ambiguous nature of the stage characters in their chaotic, topsy-turvy world. The popular reading of pantomime defines it as a fantastical escape in which harmlessly fun, stock characters act out traditional fairy stories in a good-humoured display of slapstick, popular song, romance and spectacle. Gyles Brandreth, for instance, claims that 'pantomime works' because it provides 'the security of known relationships ... you know where you are: good conquers evil, boy meets girl'.[1] This limited reading fails to analyse the inherent contradictions of pantomime where all is *not* what it seems.

A second reading employs carnival theory to interpret pantomime as a necessary and ritualistic inversion of social hierarchy and gender.[2] The pantomime's use of masks, transvestism, physical jokes and tricks, sweets and goodies for the children and a temporary relaxation of class boundaries would seem to be the British way of periodically breaking its own social rules in the safe enclosure of theatre. The cross-dressing of pantomime becomes a necessity in a patriarchal society in order to safely manage imposed gender roles. When viewed as an aspect of the carnivalesque, pantomime becomes an arena for the replacement of order by chaos, an inversion of social conventionality and a celebration of the repressed. The humour of the pantomime dame is of contrast and inappropriate behaviour in relation to the ideological expectations of the female. Thus one way of viewing the total irreverence of carnival is as politically challenging, mocking authority and solemnity. But when Victorian pantomime is contextualised, it acts as a reinforcement of the very attitudes it parodies – an acceptance of the prevailing hegemony.

A further reading psychologically analyses the ritualistic tensions expressed by the carnivalesque aspects of pantomime. Myths enacted through pantomime reflect deep-rooted anxieties and fears that are played out on stage by characters representing aspects of subconscious complexes. David Mayer proposes that the fantasy of pantomime is a mechanism for dealing indirectly with disguised anxieties. Although pantomime does not directly confront reality, he states that it 'answers the psychological realities generated by events in the tangible, verifiable world'.[3] In his article 'The Sexuality of Pantomime' he presents a Freudian analysis of the principal pantomime roles.[4] Freud interprets the spectacle of the cross-dressed man as a re-enactment of the mother who possesses a penis, therefore denying the knowledge of her castration and the subsequent anxiety it provokes. The Dame – the phallic mother – becomes a means of indirectly confronting and defusing the male castration complex.[5]

Roger Baker, in his analysis of drag performance, applies not only the Freudian theory of the castration complex, but also extends to the theory of the 'vagina dentata'. The mocking and parody of the post-menopausal woman or mother figure through cross-dressing satisfies the need to combat the fear of castration; although the small boy fears the difference between himself and the mother, he envies the mother's ability to nourish him from her body. He therefore objectifies and fetishises parts of the female body and clothing that belongs to it:

> When a boy first discovers that a woman has no penis the sight raises for him fears of his own possible castration, hence the concept of the 'vagina dentata' – the vagina that both bites and eats. When the drag queen unveils, or is unveiled, the man is confronted not with an absence but with a significant presence.[6]

Mayer, Senelick (in *The Changing Room*) and Baker, all emphasise the presence of the phallus.[7] The humour of the pantomime Dame then, according to this group of writers, relies on the tension of the female who clearly has a penis. Although considered an essential and humorous stock role, the Dame clearly embodies a complex theatrical tradition of gender play and sexual psychology.

To date, as the above examples demonstrate, all academic writing interprets the Dame's role from the male spectator's phallogocentric view, none of them considering the psychological reception of the female or that their emphasis on the phallus does not take into account a hugely relevant body of feminist theory. For example, Sandra M. Gilbert's and Susan Gubar's examination of the representation of women as fiends, monsters, witches or madwomen,[8] offers an alternative to the Freudian phallic argument and is only briefly touched upon by Mayer and Senelick. Mayer, however, in a more feminist vein, does argue that the dame's role sanctions acts of aggression towards the female, permitted within the controlling parameters of female impersonation. He is also the sole writer to place his theories within the context of pantomime action. I am strongly aware that my own analysis relies exclusively on the male reception of the Dame – all the historical accounts of the Victorian Dame act are by male writers; I have found none written by women – the silent voices of Victorian theatrical criticism.

This chapter explores how Dan Leno – generally viewed as the archetypal pantomime Dame – both conformed to and transcended the stock role that had evolved by the end of the nineteenth century (Figure 7.1). Dan Leno was, arguably, the most popular male performer of Victorian music hall and pantomime. Coming from a family of travelling players, Leno was immersed in a variety of popular performance styles and techniques – burlesque, dance, comic song and pantomime. Leno's earlier performances could be described as 'authentically' traditional and lower-class, yet his later ones, in music-hall, pantomime and musical comedy, appealed to a wider class range, bowing to the hegemonic influences of the popular theatre's move towards legitimacy. In 1885, Leno achieved great acclaim in the London music-halls, and in 1886 he was booked by the Surrey Theatre to play Dame Durden in *Jack and the Beanstalk.* Coinciding with the rise of music-hall performers taking the leading roles in pantomime, he attracted the attention of Augustus Harris and, after fulfilling a further contract at the Surrey, between 1888 and 1903 he starred in an unbroken succession of pantomimes at Drury Lane, terminated by his death in 1904. In theatrical histories he is, perhaps, still celebrated mainly as a pantomimist, eclipsing his equivalent popularity and fame in music hall.

One of the first opportunities for suspending one's habitual classification of gender roles is the first visit to the Christmas pantomime – traditionally the child's first experience of live theatre. Hugh Walpole encapsulated the sexual ambiguity of the pantomime Dame through a boy's confused reading of the stereotyped old woman, witnessed in the early 1890s:

> In the middle of the scene was a funny old woman, her hat tumbling off her head, her shabby skirt dragging, large boots and a red nose. It was from this strange creature that the deep ugly voice proceeded. She had, this old woman, a number of bales of cloth under her arms, and she tried to carry them all, but one slipped, and then

Figure 7.1 Dan Leno as pantomime dame

> another, and then another; she bent to pick them up and her hat fell off; she turned for her hat and all the bales tumbled together. Jeremy began to laugh – everyone laughed; the strange voice came again and again, lamenting, bewailing, she had secured one bale, a smile of cautious triumph began to spread over her ugly face, then the bales all fell again, and once more she was on her knees. It was then that her voice or some movement brought to Jeremy's eyes so vividly the figure of their old gardener, Jordan, that he turned to Uncle Samuel and exclaimed convulsively: 'Why, she's a man!'[9]

Conversely, the effect produced on another boy witnessing his first pantomime principal boy is far from comic. He sees immediately that she is a woman and falls under the spell of her heightened sexuality:

> I did not think it possible that such feminine charms existed as were displayed by the Principal Boy. Ample-bosomed, small-waisted and with thighs – oh, such thighs! – thighs that shone and glittered in the different coloured silk tights in which she continually appeared. How she strode about the stage, proud and dominant, smacking those rounded limbs with a riding crop! At every smack, a fresh dart was shot into the heart of at least one young adorer.[10]

Rather than leading children to question gender performativity, pantomime reinforces interpolated gender roles through its contrasted representations of the younger and older woman – the misogynistic representation of the dame and the sympathetic sexualisation of the principal boy.

By the end of the nineteenth century there was increasing distaste at seeing a man realistically portraying a woman, concurrent with a growing condemnation of 'homosexuality',[11] and this invariably influenced theatre criticism. As a result, the male actor dressed as a woman was frequently classed as an anomaly. William F. Sage, writing in 1889, stated that 'a man in female garb is apt to appear awkward and ungainly and in a word 'unsexed'.[12] It became necessary for a man, when impersonating a woman, to assume a grotesque disguise to avoid any hint of ambiguous sexual implication. A man dressed as an attractive woman offered a threat to an audience, leading them to question their sexuality when the disguise was thrown off. Writing in 1898, Max Beerbohm, like Sage, was also discomforted by female impersonation whilst enjoying the aesthetics of male impersonation:

> It often happens that men impersonate women on the stage with great success. In gesture and manner, Mr Dan Leno talking about 'Mrs Kelly' is as feminine as anything could be. There is plenty of evidence that the actors who, in the Greek and in the Elizabethan drama, represented female characters, acquitted themselves well and produced the necessary illusion. Some of the most successful lawyers, doctors and merchants of our time won early laurels as Alcestis, or Antigone, or Electra, on the classic boards of the A.D.C. I myself have seen Mr Arthur Roberts and Mr Penley playing female parts almost as well as Mr Dan Leno is playing one now. But I have never seen an actress, young or old, in any theatre, playing a man's part with any verisimilitude. This fact is the more remarkable, in that acting is the one art in which women can rival men. Every drama must contain male and female characters, and, for a reason which I shall anon suggest, it is better that the female characters be acted by women. Thus, in theatrical art, women have the *locus standi* and can attain success. But the fact remains that, though many actors can successfully obscure their sex, no actress can ever obscure hers. The explanation is simple. Men and women are not two creatures of a wholly distinct composition. A man contains in himself the whole of a woman's nature, *plus* certain other qualities which make the difference between him and her. And for him to obscure these other qualities for a theatrical purpose is easier than for her to assume them. Now, in exact ratio as the man is more successful in a female character than *vice versa,* so it is, on the whole, more pleasant to see a woman in the character of a man than a man in the character of a woman. The greater the aesthetic illusion, the more strongly does our natural sense of fitness rebel against the travesty of nature. To me, and doubtless to most people, there is something rather uncanny, not very pleasant, in (for example) Mr Dan Leno impersonating Mrs Kelly's friend. On the other hand, when Miss Millard masquerades, in *Lady Ursula,* as a youth of the last century, the effect is simply pleasing. The bounds of sex remain inviolate.[13]

Although Beerbohm makes a distinction between serious and parodic female impersonation, he views the travesties from within Victorian patriarchal society. From a modern perspective, Leno's impression of 'Mrs Kelly's friend' presents a particularly *defeminised,* unconvincing travesty. Paradoxically, although Beerbohm finds Leno a *convincing* woman in his speech and manner, he is sexually discomforted by the impression. In Judith Butler's terms, Leno's gender parody displaces the 'normal', confounding the stability between the accepted roles of the masculine and the feminine. Leno's 'fe-male' performances could be viewed as subverting 'the naturalised categories of identity and desire' presenting a challenge to the nineteenth-century heterosexual ideal.[14]

Therefore, in order to control and compartmentalise any possible subversion, the Victorians appear to have drawn a clear line between the Victorian female impersonator and the burlesque portrayal of women. The female impersonator assumed a convincing female disguise, whereas the burlesque impersonator travestied the woman in the parodic Dame tradition.[15] Leno's early Dame roles replicated a series of grim, lower-class, violent, domestic parodies played in burlesque theatres and in the music halls. The disguise of the post-menopausal, sexually inactive, ugly old woman had become the standard object of parody, offering minimal opposition to Victorian gender ideology. Leno's Dame characterisations relied on a well-established burlesque tradition of female parody for which accusations of transvestism or homosexuality had not yet been articulated. The Dame parody was controlled, partly to support the ideological construction of the female and partly to avoid discomforting men by triggering latent homophobic or homosexual instincts.

Leno's Drury Lane character, Sister Anne, from *Bluebeard* (1901) embodies all of the Dame's negative female attributes (Figure 7.2). Perrault's story of Bluebeard can be read in many ways. Viewing Bluebeard objectively, he appears as a misogynistic idealist who murders the women who disobey him, but he has also been interpreted as representative of the male psyche – his secret, dark, interior world can only be saved by the freshness of a new love that will remain discreet and not seek to delve into his past.[16] In this interpretation, the woman who trespasses through the forbidden door represents woman's curiosity and untrustworthiness, condemned for her actions by an apparently loving but betrayed husband. Bluebeard, then, becomes an easy vehicle for the standard ridiculing of the nosy wife, so favoured by Victorian humorists. Whilst Bluebeard survives almost without criticism of his cruel and bullying actions, Sister Anne is constantly ridiculed for setting her sights on a rich and powerful husband. Although the principal girl, Fatima, represents Bluebeard's attractive but curious wife, it is Sister Anne who becomes the means of entering the forbidden chamber, thus confirming woman's insatiable curiosity and insecurity about the past.[17] In Sister Anne we find the archetypal Victorian spinster. J. Hickory Wood (the author of the pantomime) exaggerates the humour by providing Bluebeard with six previous wives, described as the 'old have-beens', who, in their decapitated state, are now 'on the shelf', contrasting them

Figure 7.2 Dan Leno as Sister Anne in *Bluebeard*, Drury Lane 1901

with six youthful, beautiful wives, who enchant Bluebeard with their winning femininity. The forty-year-old Sister Anne is an ugly parody of this ideal, her vanities and pride are laughed at when she tries to pass herself off as one of the beautiful wives.

Sister Anne conforms to Mayer's analysis. Her parody of the 'fastidious middle-aged woman', with an 'inslakeable and indiscriminate sexual appetite' and a 'meekness that is improbably paired with determination to dominate', sanction Bluebeard's acts of aggression towards her. But she also 'invites momentary sympathy and insight', as experienced by Wood when he witnessed Leno playing Sister Anne:

> The more Fatima repulsed Bluebeard, the more tenaciously did Sister Anne cling to him. She saw that Fatima was preferred to her, but she could not understand why; and when, after dancing before Bluebeard to show him that she could charm as sweetly as any Fatima, she was rudely told that she was 'as graceful as a steam-roller,' her pitiful droop, as she sadly communed with herself, 'I wonder if I push myself forward too much,' was a study in sentiment.[18]

Leno's ability to bring each character to life won him unanimous approval. No matter how aggressively misogynist, Leno's art of pathos and character study managed to subtly differentiate between what could easily have become stock dame roles, endearing him to even the severest of critics. Bernard Shaw, for example, who generally expressed little interest in pantomime, referred to Leno's 'genius' and 'delicate *nuances*'.[19]

The character of Sister Anne was designed to appeal to the dominant middle-class hegemony favoured by Drury Lane, distancing her from the earlier, violent, lower-class roles played by Leno. Sister Anne's frequently quoted harp scene combined the musical genre of the parlour song with Sister Anne's hilarious and inextricable slapstick entanglement with her harp strings. Leno's choice of the harp parodied the feminine wife; the harp becomes emblematic of the 'angel-in-the house', the parlour song strengthening the middle-class domestic vision. Leno's rendition of the Victorian home-making, domestic female contributed to a completeness of role-play that audiences had not experienced previously in performances of dame roles.

As well as her role as spinster, the Dame was often portrayed as a poor, tired, hard-working mother, devoted to making ends meet for her fatherless children. Leno frequently relied on this arousal of sympathy, inviting the audience to share intimate, 'womanly' confidences. He enticed the spectator to forget the Dame's essential defence – that she is a man underneath it all; Clement Scott remarked that, 'when we see Dan Leno as a woman and hear his delightful patter it never strikes us that he is a man imitating a woman'.[20] This experience, of Leno as the convincing female, is confirmed in the article discussed earlier by Max Beerbohm. Beerbohm's emotional response however, unlike Scott's, is to be discomforted by female impersonation. Although Leno's grotesque parody of a 'slavey' such as 'Mrs. Kelly' did not contradict Scott's own perception of the working-class woman,[21] another way of interpreting Scott's reaction is to refer to what Beerbohm described as 'the woman's nature' within the man: the 'female' emotional responses to cross-dressing. Froma I. Zeitlin, in an essay on the feminine in Greek drama, argues that emotions are gender-identified and that 'playing the other' (or cross-dressing) opens the masculine self to usually banned 'female' emotions such as fear and pity.[22] Man acquires a double consciousness by dressing as a woman, enabling him to experience the feminine. When Leno created moments of real tragedy and pathos through a character such as Sister Anne, he appealed to the feminine in men such as Scott and Wood by enabling them temporarily to forget the parodic cross-dressing and experience their own interpretation of the feminine. In the pantomime *Mother Goose,* Leno sought this deepened emotional realism in his role as the female, moving away from traditional parody.

Mother Goose (1902) was considered to be Leno's triumph (Figure 7.3). In *Mother Goose,* Wood consolidated all of Leno's previous Dames to create the archetypal Dame, whose influence can still be found in pantomime today. Wood invented a new story in which the Demon challenges the Fairy to reveal to him a man or woman who can be said to be content. The Fairy seeks to prove that Mother Goose, although poor, is happy; the Demon sets temptation in her path in an attempt to prove that it is only the ignorance of riches that allows Mother Goose to remain in her state of contentment. Mother Goose experiences a series of transformations – typically pantomimic. Although still old and plain, she finds wealth through her goose's golden eggs; she is then transformed into the young and beautiful Mother Goose; finally being restored back into the original old Mother Goose. Wood stated that in all these phases Leno's characterisation of Mother Goose was 'consistency itself'.[23] Leno evoked sympathy and pity, alternating with hilarity and parody:

> When she was poor, she was humble and unassuming.
>
> When she was wealthy, she had the 'grand air' obviously put on with her riches.

Figure 7.3 Dan Leno as Mother Goose, Drury Lane 1902

> When she was young and beautiful, she was as coy and skittish a young creature as one could wish to see.
>
> And, when she returned to her pristine state, it was with a sigh of relief that she welcomed back her old familiar 'top-knot' and even her bunion.
>
> At times screamingly and irresponsibly funny, he treated the part where it required to be thus treated quite seriously.
>
> When Mother Goose first discovered that wealth without beauty was not everything, and lamented that her face was sufficient to bar her ever marrying an aristocratic title, and, again, when she found out that beauty without wealth is even a greater drug in the matrimonial market, it was no uncommon thing to hear people in the audience, in the intervals of spasmodic laughter, remark with feeling, 'Poor old soul! She makes me feel quite sorry for her.'[24]

Mother Goose was an exemplar of Victorian/Edwardian social ideology – not only was it impossible for her to rise in society without demonstrating her inherent ignorance of social mores, good taste and fashion, but beauty was inadequate without wealth and good breeding. The scene in which Mother Goose becomes wealthy turns her into a vehicle for the physical insults levelled at the pantomime dame. But the later rejection of the young, beautiful Mother Goose represents something deeper than the standard lampooning of the poor, ugly widow. Although Wood had a female stereotype in mind 'golden hair, arched eyebrows, big hips, small waist, etc. etc.',[25] Leno appears to have sought seriously to represent her physically as an innocent young girl. Although his make-up in photographs appears heavy and exaggerated the impression from the vast auditorium would have been convincing, a move towards serious female impersonation. He does not choose to glamorise her, but dresses her in respectable, cheerfully gauche lace and flounces with innocent little bouquets and floral decorations.

In a painting executed during a mental breakdown in 1903, Leno depicted himself with two heads, one the old Mother Goose and the other described in the catalogue of his paintings as 'blonde and female'. He signed it 'Yours Truly Mother Goose Dan Leno June 7th'. At the top of the sheet he wrote '60 Years of Age time changing 6 minutes to 16 years of age'.[26] It is difficult to ignore the evidence suggested by this small watercolour. Leno was very mentally disturbed during this period – in this picture he depicts himself as a split personality, the old and the young Mother Goose. It was frequently stated that Leno actually appeared to 'live' his parts both in pantomime and the music hall. Leno appears to have *lived* the roles of the young and old Mother Goose on stage, playing them as two facets of the same character, with both roles becoming mentally inseparable from himself. He played the old Mother Goose as a humble, sixty year old lady and the young Mother Goose as a skittish sixteen year old girl – not as a man playing a woman. In the painting he is also describing the physical process of the quick-change from one costume and make-up to the other. With years of experience behind him, he was proud of the short time it took to complete a total physical transformation. He is looking back to his professional triumph as Mother Goose. It was his own belief in and identification with her that led others to believe in her too, eliciting genuine feelings and responses from his audiences.

But Leno's genuine characterisation is thwarted by Wood's script, unable to avoid the conventions of the stock pantomime dame. The script takes a strange turn, for it is not only the potential suitor who rejects the beautiful, young Mother Goose but also her own son, on the grounds that she is 'distasteful'. The script suggests that Mother Goose's loss of wealth is responsible for her suitors' rejection of her; but both the suitors' and her own son's disgust at her transformation into a beautiful young girl can be explained through Mayer's analysis of the incest taboo in relation to the Dame's role. Mayer suggests that as the middle-aged ugly woman, inept in her dealings with men, 'the dame is a means of dispelling the threat implied by the mother's sexuality':

> Men not only have to accept that their mothers may be sexually attractive to men other than their own husbands, they must accommodate anxieties aroused by their own incestuous impulses. One manner of setting aside these anxieties is to bring the mother to the stage as gauche and sexually unappealing. The very ridiculousness of the dame is a guarantee that she is sexually unattractive, and will consequently inhibit guilty thoughts of incest.[27]

Leno did not want the young Mother Goose to appear ridiculous. By portraying her as a real woman he managed to incite sympathy. By creating a young and beautiful dame (the exception to the rule) and due to Leno's way of playing her 'straight', Wood, in this instance, was unable to target the usual derision at the dame's physical deterioration and sexual hunger; as a beautiful young woman there was no reason for her to be insulted. Thus Wood turned to another psychologically prompted reason for abuse by criticising her role as mother.

The young Mother Goose appears as a beautiful water nymph, discovered by her own son Jack, who does not recognise her as his mother. The spectator is required to suspend disbelief, accepting the scene as a dialogue between mother and son. Jack is attracted to the young Mother Goose, flirting with her and drawing the other men's attention to her appealing sexuality. But when he discovers that she is his own mother he is disgusted by her, hiding his guilt at his own incestuous feelings. Mother Goose is suddenly viewed as a monster, 'a pretty sight'; she is attacked for her desire to emulate a young, sexually viable girl 'at her time of life'. Jack cannot accept that his own mother might be sexually attractive either to himself or to other men. Once again the dame's role in confirming Victorian ideology is reiterated – a woman past her prime is only worthy of men's ridicule and disgust and must conform to a suitable role as mother. So, when Wood stated that it was with a sigh of relief that the audience welcomed back the old Mother Goose it was because both Leno and Wood had strayed into a world forbidden to the dominant moral ideology. Pantomime, even with its gender reversals and titillating innuendoes and costumes, served to confirm hegemony.

Although most critics appear to have received Leno's characterisations positively, there were still many who found the idea of a man in women's clothing vulgar and unnecessary – they felt that there was something intrinsically or morally wrong about one gender dressing up as another. The main criticism of Leno in pantomime was rooted in prejudice against the music hall. Critics resented the 'invasion' of the pantomime by low comedians and music hall stars.[28] But Leno appears to have survived his pantomime career with relatively little criticism. Max Beerbohm went so far as to suggest that whilst 'neither refined nor pretty', under the 'irridescent wing' of Leno ('that kind of genius which silences hostility and banishes doubt'), the Drury Lane pantomime was beyond reproach.[29] Nonetheless, one of the main objections to the pantomime dame was the gender reversal, which was found distasteful by many. A frequently cited critic of the late Victorian pantomime, W. Davenport Adams, writing in 1882 on 'The Decline of Pantomime', demonstrates objections to both these aspects of late-nineteenth-century pantomime. Adams attributed pantomime's decline to the music hall element among performers. He also criticised the travesty element of pantomime, attributing it to the influence of burlesque. He not only objected to the man dressed as a woman, but also to the principal boy in tights – clearly intended as a sexual attraction in what was ostensibly a children's entertainment. He viewed all gender reversal as unnecessary: 'A man in woman's clothes cannot but be more or less vulgar'.[30]

Yet there appears to be no criticism of Leno in this respect. Critics are unanimously eager to defend his dame roles:

> Dan Leno on the stage was seen at his best in depicting female characters ... For in Leno's theatrical performances there was nothing of offence, and he could wear the clothes of a pantomime dame in the most unobjectionable way.[31]

In *Cinderella* (1895),[32] Harris broke with pantomime tradition by using actresses as the ugly sisters. The critic of the *Sketch* warmly approved of this move but, although in favour of eliminating cross-dressing from the pantomime, he too made an exception for Leno:

> Sir Augustus Harris has struck out in the right direction in eliminating some of the vulgarities which were once thought inseparable from pantomime. Why it should be considered funny to put a man in woman's clothes it is difficult to understand. Sir Augustus Harris, like Mr. Barrett has not yet quite abandoned the fashion, yet, in retaining only one man in feminine attire, he has reduced the objection to the least possible dimensions in the circumstances. Fortunately, Mr. Leno, as the Baroness, is not repellent or vulgar, although one of the songs might with advantage be expunged.[33]

Leno won his audience's approval as a pantomime dame for a number of reasons. In his early pantomimes, his skill as a burlesque performer, incorporating parody, dancing, song and energetic slapstick, won initial recognition. As time went on and writers and directors began to shape the pantomimes around his own style of performance, Leno was allowed to develop his acting ability, creating believable characters:

> To Dan Leno a pantomime character was a real character; a character to be as carefully studied and thought out as that of the emotionally complex nature of a heroine in a problem play.[34]

For Leno, Drury Lane was the pinnacle of his career, as high as a low comedian could get in the legitimate theatre. Actors and actresses before him had revered Drury Lane as the home of legitimate drama.[35] The awe with which Leno regarded Drury Lane is well-illustrated by an anecdote cited by Wood, recounted with myth-making theatricality:

> When Dan first came back to London, he had, curiously enough, never seen Drury Lane Theatre at all; so one morning Danvers said

to him, 'Dan, I've often heard you say you've never seen "The Lane." Let's walk over and have a look at the outside now.'

So they walked over and paused before its portals and classic pillars. After a few moments of silent survey, Dan Leno, to the surprise of Danvers, walked up the steps and knelt on the topmost one. Then, rising and returning, he took his chum by the arm and remarked very quietly, 'Johnny! I shall act there one day.'[36]

Notes

1. Gyles Brandreth, 'Oh yes it is!' *Sunday Telegraph Review* December 10, 2000, pp. 1–2.
2. There has to date been little academic analysis offered on pantomime as a whole; one has to turn to general histories such as Gerald Frow, *Oh, Yes it is!* (London: BBC Books, 1985) and Derek Salberg, *Once Upon a Pantomime,* (Luton: Cortney Publications, 1981). Both speculate as to the origins of pantomime, but neither traces a connection to carnival. For an analysis of carnival see Michael Bahktin, *Rabelais and his World* (Cambridge, MA: MIT Press, 1968); Peter Stallybrass and Allon White, *The Politics and Poetics of Transgression* (London: Methuen, 1986); Victor I. Stoichita and Anna Maria Coderch, *Goya The Last Carnival* (London: Reaktion Books, 1999).
3. David Mayer, 'The Sexuality of Pantomime', *Theatre Quarterly* 4.13 (March/April, 1974), pp. 55–64, p. 56.
4. Mayer, 'The Sexuality of Pantomime', p. 56.
5. Ibid., p. 60.
6. Roger Baker, *Drag* (London: Cassell, 1994), p. 128.
7. Laurence Senelick, *The Changing Room: Sex, Drag and the Theatre* (London: Routledge, 2000), pp. 228–57.
8. See Sandra M. Gilbert and Susan Gubar, *The Madwoman in the Attic: The Woman Writer and the Nineteenth-Century Literary Imagination,* 2nd edition (New Haven and London: Yale University Press, 2000).
9. Hugh Walpole, *Jeremy* (London: Cassell, 1919), pp. 74–5.
10. Ernest Shepard, *Drawn from Memory* (London: Methuen, 1971), p. 186. First published in 1957, this is a confused, inaccurate, but evocative account of the 1888–9 Drury Lane *Babes in the Wood.*
11. Baker, *Drag,* p. 156, states that the term 'homosexual' was not coined until 1869 and not current in England until 1890.
12. William F. Sage, 'Impersonators of Women', *Theatre* vol. V, no. 13 (1889), pp. 284–6.
13. Max Beerbohm, 'Max, Mr Archer, and Others', 15 October 1898, *More Theatres 1898–1903,* ed. Rupert Hart-Davies (London: Rupert Hart-Davies, 1969), pp. 60–1.
14. See Judith Butler, *Gender Trouble: Feminism and the Subversion of Identity* (New York and London: Routledge, 1990) cited in Vincent B. Leitch (ed.), *The Norton Anthology of Theory and Criticism,* (New York and London: W.W. Norton & Company, 2001), pp. 2485–501.

15. See Senelick, *The Changing Room,* pp. 295–325, for examples of female impersonators.
16. David Johnson, *Bartok: Bluebeard's Castle* (sleeve notes for Sony disc, 1976).
17. A Biograph film was made of Sister Anne conversing with the murdered wives in the Chamber of Curiosities. This was the only film to show Leno in one of his Drury Lane roles, but no longer survives.
18. J. Hickory Wood, *Dan Leno* (London: Methuen, 1905), p. 133.
19. George Bernard Shaw, 'London Music', 23 January 1897.
20. Quoted in Baker, *Drag,* p. 173.
21. Leno's basic costuming of the dame was identical to that of the female characters in his early music hall songs – 'a plain skirt, an apron or shawl, sturdy boots and her hair parted severely in the middle, drawn ruthlessly back behind the ears and decorated perhaps with a top-knot or stray ringlet'. See Baker, *Drag,* p. 177.
22. Froma I. Zeitlin, 'Playing the other: theater, theatricality, and the feminine in Greek drama', *Representations* II (summer 1985), pp. 63–94, cited in Lesley Ferris, ed., *Crossing the Stage: Controversies on Cross-Dressing* (London: Routledge, 1993), p. 12.
23. Wood, *Dan Leno,* p. 135.
24. Ibid.
25. J. Hickory Wood, *Mother Goose,* Sc. VII.
26. Peter Wood, Catalogue for 'An important collection of original watercolours painted by Dan Leno', n.d., Cambridge.
27. Mayer, 'The Sexuality of Pantomime', p. 60.
28. See A. E. Wilson, *Pantomime Pageant* (London: Stanley, Paul and Co., 1946), pp. 57–61.
29. Max Beerbohm, 'Pantomime for Children', 14 January 1905, in *Last Theatres 1904–1910* (London: Rupert Hart-Davis, 1970), pp. 116–17.
30. W. Davenport Adams, 'The Decline of the Pantomime', *The Theatre* n.s. Vol. 4 (February 1882), pp. 85–90.
31. *Daily Telegraph,* Obituary, 1 November 1904, p. 7.
32. Augustus Harris, Cecil Raleigh and Arthur Sturgess, *Cinderella,* 1895.
33. *Sketch,* 22 January 1896.
34. Wood, *Dan Leno,* p. 119.
35. See, for example, Jacky Bratton, *New Readings in Theatre History* (Cambridge: Cambridge University Press, 2003), pp. 128–9, for the actress Fanny Kelly's reminiscences on Drury Lane.
36. Wood, *Dan Leno,* p. 117.

Part 3
Provincial Pantomime

8
Mapping the Place of Pantomime in a Victorian Town

Jo Robinson

Introduction

On Boxing Night 1865, Nottingham's New Theatre Royal – which had opened in September that same year – staged its first pantomime, *The House that Jack Built*. 'The house was filled in all parts to overflowing, more than 2000 being present' according to the report in the next day's *Nottingham Journal*.[1] And what greeted those spectators, as the preview to the pantomime in the *Journal* made very clear, was a production that was very precisely located in Nottingham itself: 'the librettist has interwoven with it [the well-known and popular nursery story] certain references and remarks which give a local habitation and a name'.[2]

The name of the pantomime itself allowed for a direct reference to the two brothers, John and William Lambert, local lace manufacturers and councillors, who had funded the building of the new theatre: 'The words, "the House that Jack (and Will) built" , have been made to apply to the new temple of Thespis which has been so recently erected in our midst'.[3] 'When the names of Messrs. John and William Lambert were mentioned by Miss May Travers from her Fairy Bower', reported the *Journal* the next day, 'there was a perfect furore, which both gentlemen gracefully responded to by bowing their acknowledgements from the box'.[4]

As Jill Sullivan's chapter in this volume makes clear, such local and topical references are far from unusual in provincial pantomime: her work on Nottingham's pantomimes, including this one of 1865, highlights the role of pantomime in the nineteenth-century town as commenting on and conversing with its audience about key issues of the time.[5] This chapter, however, takes a different approach to pantomime's relationship with its local audience; an approach that is prompted by another key moment of performance in this 1865 production. For the 'habitation' highlighted by the *Nottingham Journal* critic in his preview of the pantomime was not created through verbal allusions alone, but also through a material act of staging which emphasised the geographical and architectural space and place of the new theatre to its festive audience. Another scene which 'greatly amused', according to the *Journal*, was:

> The erection of 'The House that Jack Built,' by a hundred juvenile masons, bricklayers, and carpenters, who speedily erected 'The House that Jack (and Will) Built', in the shape of an excellent scenic reproduction of the front of the new theatre.[6]

That front of the new theatre – with its 'portico, supported by six elegant Corinthian columns' of Ancaster and Mansfield stone – had excited interest at the time of the opening of the theatre in September 1865, when the architect C. J. Phipps noted that 'the style of the façade was fixed by his clients, and not suggested by him; consequently whatever merit attached to it must be awarded to them'.[7] The creation of this façade on the stage on this first night of the pantomime season, by one hundred Nottingham children and in front of those clients who had elected for such a grand design, can thus perhaps be situated as a moment in which stage looked back at audience, pantomime to its community, and shared in a moment of celebration of this new, grand, civic architecture which had established a new landmark in the town. On Boxing Night and for the next six weeks from then until the closure of the full pantomime on 3 February 1866, the audience of Nottingham's New Theatre Royal shared in a moment of performance in which the building of the new theatre was highlighted and celebrated on the stage of that theatre, reflecting the pride of the builders and their audience back to that audience from the stage of the theatre.[8] Taking its cue from this celebratory imaging of the theatre to the audiences for this pantomime, this chapter thus seeks to explore the place of pantomime within the town of Nottingham in the mid-nineteenth century.[9]

Mapping place/mapping connections

In *New Readings in Theatre History*, Jacky Bratton argues that any individual performance is part of a continuously shifting network of relationships – 'a collaboration in the creation of a particular theatrical experience, taking place first in writing, casting, rehearsal and designing and then anew, afresh each night in front of a new audience':

> The single night in the theatre is a point of crystallisation in a continually moving, dissolving and re-forming pattern, most elements of which are [...] not only unrecorded but unremarked, though not uninfluential.[10]

Where Bratton's intertheatrical focus is largely on the creators of performances and performance texts – playwrights, actors, designers, managers – this chapter widens the scope of attention to examine also the relationships between the places and spaces of performance and the spectators who attended them. It is here, in thinking about the connections between performance, place and audience that the idea of mapping proves particularly useful. For, according to David Turnbull, maps themselves prompt a consideration of the relationships *between* things, moments, places:

> At base there is something more than merely metaphoric about maps and theories; they share a common characteristic which is the very condition for the possibility of knowledge or experience – connectivity. Since we cannot have a pure unmediated experience of our environment, that experience is better understood as an active construction resulting from a dialectical interaction between the 'lumps' in the landscape and our imposed connections of those lumps. Our experience and our representations are formative of each other and are only separable analytically.[11]

Mapping the place of pantomime necessitates, then, a focus not just on the pantomime itself, and on its performance on the stage of the new theatre, but also on the series of relationships between that annual performance and other elements in the town: between pantomime and a competing range of entertainments associated with the Christmas period; between the pantomime stage and the variety of audiences and individual spectators who attended, or were urged to attend through playbills and advertising; and, as highlighted through the performative creation of the very place of pantomime on Boxing Night 1865, between this venue and others, this place and other locales, within the town. As a long running production – lasting from Boxing Night into

February in a successful year, such as this one – pantomime offers the theatre historian the unique opportunity of an extended, and detailed, study of such relationships, enabling us to explore the place of theatre within the society of a particular town.

Mapping the place and space of pantomime

Turnbull's claim quoted above, that 'our experience and our representations are formative of each other', seems to echo the ideas of Henri Lefebvre in *The Production of Space,* where he argues for a triadic understanding of space as always and simultaneously being perceived, conceived, and lived.[12] Working with a map of the centre of Nottingham which lays out the sites of performance in geographical relation to each other, this section highlights not just the spatial position of the Nottingham theatres and other sites of entertainment, but seeks to analyse those sites in terms of what Lefebvre calls their 'representational spaces': that is, spaces as they are directly *lived* through their associated images and symbols, and hence the space of 'inhabitants' and users.[13] According to Lefebvre, representational space 'overlays physical spaces, making symbolic use of its objects'.[14]

When the New Theatre Royal opened in September 1865 it was, as its name of course implies, not the first Theatre Royal in Nottingham. The old Theatre Royal, built in 1760, had finally closed in March 1865 (although the building was to be rapidly refurbished and re-opened as Middleton's Alhambra Palace of Varieties only four weeks later). The positions of the two sites for the town's Theatre Royal can be seen on the map (Figure 8.1) – the first, situated away from the Market Place on St. Mary's Gate (in the far south-east corner of the map extract), the second, built for the Lambert Brothers in 1865, sited above the Market Place and connected to it by a broad street which cleared a path through what – on this map – is still shown as slum housing. However, this move in position of the Theatre Royal site across the improving town of Nottingham reflects a change not just in the physical space of the theatre, but also in the ways that the space of the theatre is conceived and lived within the nineteenth-century town.

The old Theatre had long been criticised as unsuitable for a respectable audience, described in 1852 as 'a small mean building, dirty within and without, which no one who has not seen it would be able to find out, though he searched for a week'.[15] The anonymous critic Asmodeus, writing his *Reflections on Life in Nottingham* in 1860, also expressed disgust that '*the* Theatre erected just a century ago, in St. Mary's Gate,

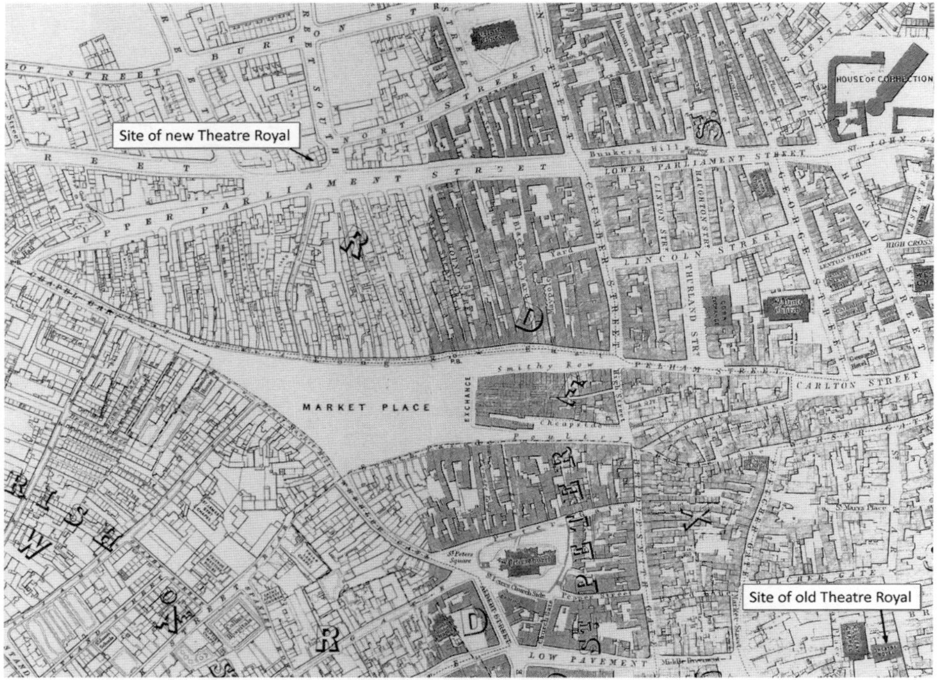

Figure 8.1 Excerpt from 1861 plan of the town of Nottingham and its environs, by Edward W. Salmon, showing the position of the two Theatre Royal sites in relation to other key sites in the town (Nottingham City Council Leisure and Culture Services Local Studies Library)

is the Theatre still!' while Kathleen Barker's later analysis of the state of theatre in Nottingham at this time suggests that

> the combination of a slum site, an old building too small and plain to be attractive for prestige entertainments, and the strength of Dissenting opposition made it an unattractive managerial proposition even in times of prosperity.[16]

Asmodeus's other comments suggest, however, that the decay of the theatre was not caused by a lack of appetite for performance within the town:

> We see in the miserable 'Hops', – the public house 'Concert-rooms', – the wretched 'Circus' – that there is an appetite for amusement which, because it cannot be gratified on the board of the regular Theatre, palls itself on the garbage and rubbish of those miserable substitutes, and drinks in suggestive profligacy and unblushing obscenity, instead of pure waters from the founts of genius and inspiration.[17]

Asmodeus's other essays – on the free and easy at the Royal Colosseum, and on the scandalous presence of 'middle-aged men, with families looking up to them for example and position', at St. George's Hall – perhaps prompt us to approach his evidence with a critical stance, but what we can glean from this brief discussion of the old Theatre Royal is that it was, by the mid-nineteenth century, too small and too plain, lacking 'prestige' in Barker's terms, and discussed by commentators not in relationship to high culture, but to the 'Hops' and 'Circus'. To return to Lefebvre, the physical space of the theatre – an eighteenth-century building no longer suitable for the developing town, and situated in a 'slum site' – was thus overlaid with a perception of the space as not respectable or fitting, inside or out.

The introduction to this chapter has already called attention to the contrasting grandeur of the New Theatre Royal building, with its 'six elegant Corinthian columns' supporting 'an imposing entablature with plain frieze and bracket cornice': it is clear that architecturally, the physical space of the new building signalled a very different cultural place for theatre within Nottingham.[18] Such a change is also echoed in the geographical location of the new theatre, sited in an area of the town which, while it itself had recently been a slum site, at the top of what one observer described as the 'long-standing nuisance of Sheep Lane' (visible on the map of 1861 which accompanies this essay), was now part of a project of improvement which was transforming both physical and representational space.[19] By the time of the opening of the theatre on 25 September 1865, 'on a scale of magnitude and elegance rivalled only by the first class theatres of Europe' (in the view of the *Express*), the building sat at the top of a 'new and spacious street leading from the theatre to the Market Place', replacing what the *Tourist's Picturesque Guide to Nottingham* of 1871 described as the 'narrow, dingy dirty thoroughfare of Sheep Lane, and the mean little cottages which [... previously] formed the boundary line upon which now stands the new Theatre and its adjuncts'.[20]

Thus far, this section has considered the place of pantomime largely in terms of the buildings in which those pantomime performances took place; indeed, the scene and the title of *The House that Jack Built* with which this chapter opens encourage such a focus. In the next section, however, I turn to examine the relationships between pantomime and other festive performances within the nineteenth-century town; with this in mind, a brief consideration of the relationships between the different sites in which such performances took place is merited. For the move of the Theatre Royal from St Mary's Gate to the new site represented, as the *Nottingham Journal*'s report of the theatre's opening made clear, a move from a decaying site to what 'may be called the new town which has sprung up beyond the boundary of line of Parliament Street, and which comprises so large a proportion of the more respectable streets and residences of the borough'.[21] Here then the new theatre was associated with both a physical improvement and increased respectability – with the possibility of prestige, in Barker's terms. In addition, the move also brought the new Theatre closer to other sites of performance, particularly the Mechanics' Institute which was now only a few streets away, as well as the Exchange Rooms in the Market Place. Seeing these different sites connected by easy walking distance on the map of the town suggests a potential for audience choice, and competition for audience, which would be particularly acute during the long pantomime run. An examination of synchronous performance events thus superimposes a performance map over the geographical one.

Mapping performance repertoire

While the Theatre Royals, old and new, were the sole purveyors of pantomime within the town of Nottingham, that pantomime was not the only regular entertainment for the festive Christmas season; advertisements for other events in December and January make it clear that other venues and entertainments have an eye on the holiday audiences.

An attempt to map the cultural and social place – in addition to the geographical place – of pantomime requires us to revisit and rework Bratton's concept of intertheatricality, exploring the other potential performance choices available to the pantomime audience within the town. We have already seen how the building of the New Theatre Royal changed both the actual space and the perceived space of the theatre within Nottingham; here, through consideration of the 1856–57 and 1865–66 pantomime seasons, I seek to uncover that 'continually moving, dissolving and re-forming pattern' identified by Bratton, locating the cultural/social space of the pantomime through a mapping out of the performative landscape in which these two productions took place. Some of these other competing performances were also repeated ones, as much associated in the society of the town with the Christmas season as the pantomime itself, while others sought to appeal to audiences with potentially more leisure at this time of year than at others.

In the pantomime season of 1856–57, for example, the pantomime of *Whittington and his Cat* opened on Boxing Night at the old Theatre Royal in St Mary's Gate, attracting an enthusiastic and numerous audience according to the rather critical report in the *Nottingham Journal* the following week:

> The Christmas pantomime was produced for the first time on Friday evening last, before a densely crowded house, there not being a seat vacant from floor to ceiling. Of its merits we cannot speak very highly. The introductory portion is very poor, the dialogue lacking point, and the incidents not being sufficiently descriptive of the career of the hero. [...] The enthusiasm of the audiences on Friday was great indeed and the pantomime passed off in the most successful manner. It has been repeated nightly to similar assemblages, whose satisfaction has been most unequivocally expressed. There is every prospect of its having a long run.[22]

Performances of the pantomime did indeed continue until 23 January 1857, but during that run the Theatre Royal management did not have an open field in Nottingham. Instead the attractions of the pantomime were set alongside a number of other events, including: 'Wladislaw's Grand Mechanical Exhibition of the War in the Crimea', at the Mechanics' Institute, which ran from mid-December to early January; Dr Reentz's Anatomical Museum, Church Gate, at the back of the Post Office, which was announced to 'positively close on January 6th; a lecture on Christmas Carols delivered at the Exchange Hall, on the evening of New Year's Day, by Mr Coburn, Medical Superintendent of the Midland Institution for the Blind; Christmas tea parties at the New General Baptist Chapel, New Lenton, and at St John the Baptist's Church; a performance of the Messiah by the Nottingham Choral Society in the Wesleyan Congregational Free Church on Monday 12 January, 1857; Lancaster's Magnificent Panorama Of The Late Crimean War, with vocal and instrumental accompaniments, which opened for twelve days at the Exchange Rooms, Market Place, commencing Monday 12 January, 'introducing the most important Battles and startling Incidents which have occurred during the campaign'; Miss Smeeton's Second Annual Ball at the Assembly Rooms on Tuesday 13 January (with a recitation to commence at 7 o'clock, and the Ball to open at nine); another Ball, in connection with the Select Quadrille Party, on Thursday 15 January; John Farmer's annual concert on Monday 19 January at Mechanics' Hall, and finally, an opportunity to view the 'Magnificent Historical Pictures of "The Execution Of Montrose" and "The Last Sleep Of Argyle"' from the House of Lords which were on display at the Exchange Rooms, 11 till 5, and 7 till 9.[23]

Clearly not all of these entertainments are in direct competition with the pantomime, but many of them would appeal to different sections of the potential pantomime audience, as a consideration of pricing and advertising reveals. The costs of pantomime attendance are not given in the newspaper advertisements or reviews of the 1856–57 season, but three years later in the course of his criticisms of the theatre in1860 Asmodeus deplored the 'scandalously low prices of admission' to the old theatre as inevitably inviting a disrespectful audience, and suppressing any attempt at artistic excellence:

> The idea of getting a decent audience in the gallery at 3d a head! or, of one anything like select, in the pit at 6d! is preposterous! So again boxes, descending from 1s.6d and 1s. to 6d, prices which would really deter decent people from entering the theatre.[24]

Consideration of the advertising for these alternative events makes it clear that at all levels, from boxes to gallery, there were competing attractions for the potential spectator of pantomime. At Wladislaw's Grand Mechanical Exhibition, for example, an announcement in the *Journal* of 26 December which gave notice that it was the 'last week but one' of performances, advised readers that in order:

> to give all classes an opportunity of visiting the above celebrated entertainment, the admission will be REDUCED during Christmas week to HALF-PRICE! And for the convenience of visitors a MORNING

> PERFORMANCE will take place on Christmas-Box day (Friday), December the 26th, Saturday the 27th, Monday the 29th, and the Last on Wednesday, the 31st. Each commencing at half-past Two.[25]

Prices for these shows were thus halved from the usual admission of 1s.6d for front seats, 1s. second, 6d third, though reserved seats remained at 1s.; the special morning performances during Christmas week surely acknowledge the competing evening attractions of the newly staged pantomime. Lancaster's Crimean Panorama, swiftly following on from Wladislaw's representation of the same conflict, also aimed at a shared market; the newspaper column which reports on the 'exceedingly good' attendance at the pantomime's Juvenile Night also records that:

> Our juvenile friends have found no lack of amusement during this festive season. As an appropriate wind–up, we have now to notice the very spirited collection of panoramic views to be seen every afternoon and evening at the Exchange Hall, in which some of the most striking incidents of the late war are portrayed with graphic effect. The charms of music, both vocal and instrumental, lend additional *éclat* to the entertainment, which is well worthy the extensive patronage which it has already received. The infant drummer is a special favourite with the ladies.[26]

Ticket prices for Lancaster's Panorama were reserved seats 1s., second seats, 6d. The *Journal* of 30 January records a free showing to 250 children from Nottingham ragged schools, and workhouse children, a marketing device which, as we will see, was to be adopted by the managers of the new Theatre Royal in the 1865–66 pantomime season.

Musical entertainments, and those at the Mechanics' Hall – likely to attract the higher reaches of Nottingham's society – proved the most expensive attractions in this Christmas period, with John Farmer's concert, which was an annual constant throughout the mid-nineteenth century years, costing 2s. for front seats, second seats, 1s., promenade 6d. While the Nottingham Choral Society's *Messiah*, on 12 January, cost just 1s. or 6d, the town's Sacred Harmonic Society's performance of the same Oratorio, calling among others on the talents of Madame Rudersdorff, advertised as 'Prima Donna of Drury Lane and Sacred Harmonic Exeter Hall, and the Birmingham Festival (her first appearance in Nottingham)', which followed swiftly on the closure of the pantomime at the Mechanics' Hall, reflected in its ticket prices the status of the society's members and of its professional performers: reserved seats were 4s.

(Members of the Institution, 3s.); Second seats, 2s.6d (members 2s.); Promenade 1s.[27] In contrast the cheapest performances on offer during these Christmas holidays were those at Dr Reentz's Anatomical Museum, entrance 6d, which made up for its low price with the frequency of lectures, at twelve, two, four, half-past seven, and nine. 'Ladies days' were on Tuesdays and Fridays, 'NONE BUT ADULTS ADMITTED'.[28]

This brief consideration of synchronous performance events around the pantomime of 1857 suggests that in this, as in other years, the pantomime must be seen as just one element amongst a number of performance events within the town, and indeed on this occasion – in a theatre already seen as outdated by the *Nottingham Journal* by 1852 – perhaps less prestigious or successful than others. But with the shift in location of the new Theatre Royal, and the associated cultural shift towards greater respectability identified above, the manager of that new theatre, Walter Montgomery, sought to build a newer and more distinctive audience for the pantomime of 1865, raising prices accordingly (Private boxes from One to Two Guineas; Dress Circle, 2s.6d; Upper Circle, 1s.6d; Pit, 1s.; Gallery, 6d). Such an upwards shift distinguished the new theatre's patrons from those of the Alhambra or St George's Hall (which charged 6d for front stalls, and 6d (2d of which is returned in refreshments) for the body of the Hall and Gallery), instead aligning the theatre with the familiar family entertainments provided by Farmer's annual concerts and the return visit of the comic sisters Sophia and Annie to the Mechanics' Hall. An examination of the playbills for this pantomime printed in December 1865 and January 1866 thus maps out the new place of pantomime in Nottingham in relation to the new audiences which Montgomery was seeking to develop, locating the pantomime within the cultural and social landscape of the town.

Mapping the audience of 1865: advertising communities

'There was no such thing as a Victorian audience, but rather a variety of audiences, embodying a range of perspectives.'[29] Jim Davis and Victor Emeljanow provide an important reminder to the theatre historian that audiences – and the individual spectators within them – are always changing, never homogenous. That this is so is exemplified by the series of playbills printed during the run of the 1865–66 pantomime, which in their very different appeals to a variety of potential readers and spectators, work to identify and to attract different kinds of audiences for this seasonal entertainment. The pantomime itself may operate on the basis of 'a local habitation and a name', but the sustaining of a pantomime run in competition with other performance events in a provincial town such as Nottingham means that the theatre has to appeal to, and marshal, *different* habitations, areas and sectors of the town and its surrounding society.

According to Bratton, theatre playbills provide evidence 'for those most difficult and evanescent aspects of theatre history – the expectations and disposition of the audience, their personal experience of theatre'.[30] A focus on those elements of the playbills which directly address the constituency of the audience for this Nottingham pantomime demonstrates how the manager of the Theatre Royal sought to appeal to different expectations and dispositions, targeting potential audiences based on social groupings, status, and geographical location.[31]

The first claim made by these playbills, even before the pantomime opens, echoes the new 'representational space' of respectability which we have earlier identified as being claimed by the new theatre. On the theatre's opening night, 25 September 1865, the manager Walter Montgomery had argued from the stage for the potential of 'a well-conducted theatre' with 'a company of ladies and gentlemen', to be 'a teacher of the highest morality', directly answering the concerns of the anonymous Asmodeus as to the decay in morality and atmosphere in the theatre which it replaced.[32] In advance of the pantomime, and with a view to recruiting the juveniles who would create the façade of the new theatre on the pantomime stage, Montgomery placed a 'Special Notice' on the playbill of 11 December 1865:

> One hundred Children will be required for one of the Great Scenes in the Pantomime. None but respectable Persons need apply; and all applications must be BY LETTER ONLY, addressed to Mr H. F. SAVILLE, containing the height and age of the Child.

In recruiting these children, Montgomery was in fact repeating a device used earlier that autumn by Middleton's Alhambra Music Hall, which advertised in September for '100 little girls' who were to form the Infant Army of Aldershot later that season. Montgomery's stress on respectability is an addition, however, and is reinforced by the invitation to tender for provision of refreshments which accompanies it on the same bill:

> None need make application but persons of the highest character, who will give a guarantee that every article supplied shall be of the very best quality, and the attendants respectable and civil. For the comfort of the Audience, the Management suggest that Refreshment,

> from a cup of tea and a slice of bread and butter, to a sandwich and a bottle of champagne, should be obtainable, so that the comfort and economy of home may be experienced by the thousands who are expected to throng the Theatre during the run of the Pantomime.

In repeatedly stressing respectability as a vital element in audience, suppliers and performers, these notices seek to reassure potential spectators of the acceptability of this new theatre: where Asmodeus placed the old Theatre in relation to the Hops and Circus, Montgomery's invitation to tender instead places his new theatre in a relationship to the respectable 'home' of his potential audiences. And of course, recruitment of one hundred respectable children presumably attracts one hundred or more respectable families into the theatre along with them.

Once the pantomime began on Boxing Night, the playbills continued to appeal to the respectable society within the town, but also sought to widen the potential audience in other ways. As with performances of *Whittington and his Cat* in 1856–57, *The House that Jack Built* involved a number of Juvenile Nights, such as that of Wednesday 10 January 1866, when a GRAND JUVENILE NIGHT! was announced, and changes in the order of performances, and prices, made to accommodate the younger audience:

> THE GORGEOUS PANTOMIME WILL BE PLAYED FIRST
> Concluding with WHERE THERE'S A WILL THERE'S A WAY!
> ON this occasion, Children under 12 years of age, will be admitted at absolute Half-price to all parts of the House, viz. – Circle, 1s. 3d.; Boxes, 9d.; Pit, 6d.; Gallery, 3d. – To prevent discomfort to the little people, the Doors will be opened at SIX o'clock.

A series of Juvenile Nights were repeated through the run, and on the last occasion, in an echo of Lancaster's Crimean Panorama of 1857, the playbill of 24 January announced that:

> There will be one more GRAND JUVENILE NIGHT, on Wednesday next, the 31st, on which occasion, by the invitation of the Messrs. Lambert, the WORKHOUSE CHILDREN will attend.

The playbill here calls attention to the philanthropic generosity of the Lambert brothers, builders and owners of the new theatre, again reinforcing the sense of respectable morality stressed by Montgomery at the opening performance.

In contrast to this workhouse audience, whose invitation came at the end of the pantomime run, when the manager may have been struggling to fill the theatre, the playbill of 9 January 1866 recorded a charitable performance for spectators of much higher status within the town. A 'GRAND MASONIC NIGHT' was announced: 'On this occasion, the Theatre has been TAKEN by the FREEMASONS, and the proceeds of the Evening will be given to the CHARITIES OF THE TOWN'. The performance took place 'under the Patronage of his Grace the Duke of Newcastle', the Provincial Grand Master, 'and the Freemasons, who will appear in FULL MASONIC COSTUME', and with this audience in mind the playbill advised of a substantial rise in ticket prices:

> On this occasion (being for Charitable purposes), the Prices will be as follows – PRIVATE BOXES, £3 3. 0d.; OMNIBUS DITTO, £2 2s. 0d.; UPPER DITTO, £1 10s. 0d.; Tickets, RESERVED STALLS, 4s.; DRESS CIRCLE, 3s. 6d.; UPPER CIRCLE, 2s. 6d.; PIT, 2s.; GAL. 1s.

As with the change in repertoire for the Juvenile Nights, it was not just the prices that were changed: the playbill makes it clear that Montgomery, the manager of the new Theatre, was working to adapt the pantomime to the particular 'expectation and disposition' of his audience, which on this occasion, was one with which he was firmly aligned:

> Miss REINHARDT will speak an apropos PROLOGUE (and disclose the secret of Freemasonry), written for the occasion by BROTHER WALTER MONTGOMERY.

Echoing the dialogue between pantomime stage and audience constructed nightly through the creation of the new theatre's façade, this direct address from stage to spectators again claimed a shared space and place for this theatre, and this pantomime, within the society of the town. Across such occasions – juvenile nights and Masonic evenings – as well as in the Cricketers' Bespeak of 9 February, when the pantomime, having undergone 'some supervision and excision to adapt it to the requirements' of a second performance, was played, the pantomime was thus positioned not just to appeal to respectable families within the town but also to specific classes and groupings, mapping out a series of different relationships and appealing to potential spectators of very different status.[33]

The playbills for *The House that Jack Built* can also be seen to be appealing to different geographical areas and audiences. We have already seen through our analysis of the 1861 map that the site of the new Theatre Royal positioned it within an increasingly respectable part of the town, in close proximity to other key sites of performance, but over a six-week run the pantomime needed to be able to expand its audience beyond the boundaries of the town, drawing spectators into Nottingham with the help of the two railway companies: the Great Northern, whose station had opened in 1857, and the Midland, opened in 1848.

The playbill for Monday, 8 January 1866 thus advertised a 'GRAND DAY PERFORMANCE WHEN THE PANTOMIME ONLY WILL BE PLAYED, Commencing at 2.30, terminating at 4.50'. On this occasion both the Midland and Great Northern were announced to be running 'SPECIAL EXCURSION TRAINS'. While on this occasion the timing of the pantomime performance was changed, a notice in the playbill of 11 January announced instead a change in the times of the Midland Company's trains on Saturday, 13 January:

> An ADDITIONAL TRAIN, First, Second, and Third Class, will leave NOTTINGHAM for PYE BRIDGE, AT 11-45 P.M., calling at the following Stations, – Beeston, Long Eaton, Sandiacre, Stanton Gate, Ilkeston Junction, Shipley Gate, Langley Mill, Codnor Park. MARKET TICKETS taken to Nottingham by any of the Ordinary Trains during the day, will be available to return by this Train.
>
> JAMES ALLPORT, General Manager

A week later, the playbill which announced that 'NOTWITHSTANDING THE CONTINUED SUCCESS OF THE GORGEOUS PANTOMIME', it would shortly be withdrawn 'to make room for other Novelties', reminded potential spectators that 'The Great Northern and Midland Railway Companies, are now running Late Trains to all the surrounding Towns and Villages'.[34] The same notice was included on the playbill for Friday 26 January, which advertised the Benefit of Little Lafar the Clown, promising a 'GRAND DOUBLE COMPANY OF PANTOMIMISTS!', with 'Double Clowns! Double Columbines! Double Pantaloons! Double Harlequins!' Even within a growing regional town such as Nottingham, a successful pantomime needed to appeal beyond its narrow 'local habitation', making connections to potential audiences in nearby towns and villages. Fully mapping the place of pantomime in this context thus requires a widening of the focus of the map reproduced here, paying attention to the developing transport links which connected this Nottingham pantomime to wider Nottinghamshire and Derbyshire.

Conclusion

This chapter's exploration of the annual tradition of pantomime in Nottingham has examined the place of pantomime not just as a festive performance taking place on a particular stage, but as one which was actively connected to, and built links with, the existing performance and social culture of the town. The concept of mapping has been employed in two ways: in a geographical sense, to examine the space and place of the New Theatre Royal, and illustrate the sites of potential alternative performances through which the audience of *The House that Jack Built* would make their way to that theatre, and to this particular performance; and in a metaphorical sense, to draw out the network of intertheatrical connections between this pantomime and other performances, and between this pantomime and its varied audiences, which enable us to begin to map those potential spectators against the range of social and cultural contours existing within the nineteenth-century town. In analysing the findings of both kinds of mapping – each of which involves, in Turnbull's phrase, 'an active construction resulting from the dialectical interaction between the "lumps" in the landscape and our imposed connections of those lumps' – this essay has sought to place pantomime in its wider context, among a range of competing performances and entertainments from rival attractions at the Alhambra Music Hall and Mechanics Institute, religious celebrations of Christmas and a wide variety of more private performances in the town. Through this process, the annual pantomime can be seen both as a particular moment of celebration of and dialogue with the town, and as part of a wider network of activities and structures which shaped Nottingham's performance culture in the mid-nineteenth century.

Notes

1. *Nottingham Journal*, 27 December 1865, p. 2.
2. Ibid., 26 December 1865, p. 2.
3. Ibid.
4. Ibid., 27 December 1865, P. 2.
5. See 'The Business of Pantomime: Regional Productions 1865–1892', unpublished doctoral thesis, University of Nottingham, 2005.
6. *Nottingham Journal*, 27 December 1865, p. 2. The *Nottingham and Midland Counties Daily Express* offers a fuller account: 'A great number of little people occupy the stage, busily employed in the various handicrafts required in the erection of the house, and their operations make a perfect Babel of noise,

whilst a pigmy policeman, an infant apparently not more than three or four years old, patrols up and down in front of the works, ultimately collaring and taking off in custody a delinquent mason or carpenter. On the builders advance with their work; the platform upon which they stand rises until it reaches the flies, when it descends and reveals the House the Jack (and Will!) built – that is, the front of the new theatre.' *Nottingham and Midland Counties Daily Express*, 28 December, 1865, p. 2.

7. *Nottingham Review*, 29 September 1865, reproduced in Richard Iliffe and Wilfred Baguley, *Victorian Nottingham: A Story in Pictures*, 20 vols (Nottingham: Nottingham Historical Film Unit, 1970–83), vol. 7 (1972), p. 43.
8. *The House that Jack Built* gave its final full performance on 3 February 1866. However, a special notice inserted into the playbill of 6 February 1866 testifies to the ongoing power of the pantomime to attract audiences. 'In consequence of the very great desire of great numbers unable to obtain admission last week to witness the Pantomime […] be repeated several times after undergoing some supervision and excision to adapt it to the requirements of a second […] played with the Grand Dramas in the course of presentation. The Gorgeous Transformation Scene will, as usual, conclude the dramatic scenes, and this beautiful composition will be enhanced by the attraction of Miss Rosina Wright, who has been specially retained, and who will execute an entirely new and brilliant dance in it.' New Theatre Royal Nottingham Playbill, 6 February 1866, Local Studies Library, Nottingham City Council Leisure and Culture Services. It is not clear from this description whether the theatre construction scene was retained in this excised production.
9. Research for this chapter has been undertaken as part of an AHRC-funded project, 'Mapping Performance Culture: Nottingham 1857–1867', running from 2006–2009 at the University of Nottingham. In collaboration with Dr Gary Priestnall (School of Geography), and in partnership with the University Library's Department of Manuscripts and Special Collections, Nottinghamshire County Archives, Nottingham City Local Studies Library and Nottingham City Museums this project investigated performance and entertainment in Nottingham through development of an intuitive interactive map and research database which layers social, cultural and economic data onto a spatial representation of the town. www.nottingham.ac.uk/mapmoment.
10. Jacky Bratton, *New Readings in Theatre History* (Cambridge: Cambridge University Press, 2003), p. 38. See Jo Robinson, 'Mapping performance culture: locating the spectator in theatre history', *Nineteenth Century Theatre and Film*, 31.1 (2004), 3–17, for a detailed argument in support of extending Bratton's concept of intertheatricality to include spectators of performance.
11. David Turnbull, *Maps are Territories: Science is an Atlas* (Chicago: University of Chicago Press, 1993), p. 61.
12. Henri Lefebvre, *The Production of Space* (1974), trans., Donald Nicholson-Smith (Oxford: Blackwell, 1991), p. 39.
13. Lefebvre, *The Production of Space*, p. 39.
14. Ibid.
15. *Nottinghamshire Guardian*, 9 September 1852.
16. *Revelations of Life in Nottingham by the English Asmodeus* (Nottingham: C. B Truman, 'Telegraph' Office [1860]), p. 94; Kathleen M. D. Barker, 'The

Performing Arts in Five Provincial Towns, 1840–1870', unpublished doctoral thesis, University of Leicester, 1982, pp. 49–50.
17. *Revelations of Life in Nottingham*, p. 92.
18. *Nottingham Journal*, 30 September 1865, p. 2.
19. Extract from *The Tourist's Picturesque Guide to Nottingham*, 1871, reproduced in *Victorian Nottingham: A Story in Pictures*, vol. 7, pp. 60–2.
20. 'Inauguration of a New Theatre Royal in Nottingham', *Nottingham and Midland Counties Express,* 26 September 1865: 3.
21. *Nottingham Journal*, 30 September 1865, p. 2.
22. Ibid., 2 January 1857, p. 5.
23. Details of these performances are taken from weekly editions of the *Nottingham Journal* for the months of December 1856 and January 1857.
24. *Revelations of Life in Nottingham*, p. 94.
25. *Nottingham Journal*, 26 December 1856, p. 1.
26. Ibid., 16 January 1857, p. 5.
27. Ibid., 2 January 1857, p. 1.
28. Ibid., 12 December 1856, p. 4.
29. Jim Davis and Victor Emeljanow, *Reflecting the Audience: London Theatregoing, 1840–1880.* (Hatfield: University of Hertfordshire Press, 2001), p. 229.
30. Bratton, *New Readings in Theatre History*, p. 39.
31. All playbills referenced in this section are held in the Local Studies Library, Nottingham City Council Leisure and Culture Services, identified by date. Individual references are given only when the date of the playbill is not apparent within the essay.
32. Walter Montgomery's speech at the opening of the New Theatre Royal Nottingham, reported in 'Opening of the New Theatre Royal', *Nottingham and Midland Counties Daily Express*, 26 September 1865, p. 3.
33. New Theatre Royal Nottingham Playbill, 6 February 1866.
34. Ibid., 19 January 1866.

9

'Local and political hits': Allusion and Collusion in the Local Pantomime

Jill A. Sullivan

The experience of pantomime audiences in the nineteenth- and twenty-first centuries have certain similarities: they will have both loudly booed the villain and cheered the hero; applauded scenic and novel effects; awaited the traditional jokes and catchphrases ('oh no he isn't', 'it's behind you'), and allowed themselves to be conducted from the stage in choruses of strained rhyming couplets. In addition there will have been a certain level of expectation regarding topical comments, comments about national issues as well as about issues more pertinent to a local audience. The inclusion of social and political references has long been a feature of British pantomime, initially established in the influential and innovative mimed satire of the Georgian Harlequinades.[1] The repeal of the Theatre Licensing Act in 1843, which lifted restrictions on the spoken word in theatres, together with an increase in the spectacular nature of productions, contributed to a changed format for the annual pantomime at theatres throughout the country. The mimed antics of the Harlequinade were no longer the focus of productions, becoming instead a footnote to the more elaborate telling of the stories of *Robinson Crusoe, Little Red Riding Hood* or *Cinderella*, and the visual satire and topicality that had been so central to productions in the early 1800s were increasingly complemented by verbal satire and comment on the social and political developments of the day. Critical attention to the topicality of Victorian pantomime has tended to focus on the legitimate London theatres, where a necessary responsiveness to the Lord Chamberlain's directives frequently curtailed the tone and level of comments.[2] Although provincial theatre managements were also required to submit all scripts to the Examiner of Plays, neither pantomime submissions nor the enacting of directives were comprehensive in the provinces. Geographically placed beyond any impending physical presence of the Lord Chamberlain's officials, and with the licensing of the theatre building the responsibility of local magistrates rather than the Lord Chamberlain, pantomime managers and writers had far greater opportunity to freely address a range of relevant national, international and – importantly – local political, economic and social issues. Those references appealed to a diverse number of interest groups in each audience, but they were not comprised simply of random targets. Instead, the referencing in provincial pantomime could be both thematic and sustained over a period of years, addressing new developments and heated debates in local matters. At the Theatre Royal, Birmingham in the 1870s and 1880s, much of the pantomime topicality drew on very specifically framed political references. It is those references that I will focus on in this chapter, drawing on textual evidence from the book of words and newspaper reviews.

The study of pantomime naturally requires a caveat in which examination of the textual evidence is notoriously framed by the mutability of the genre: the changing nature of the script in pre-production and during the run and the unrecorded variations and new references which could be included and updated in patter songs and jokes. Frequently, the only remaining script of a Victorian pantomime is that in the book of words, a version of the script that was on sale to audiences in the theatre at the time of performance.[3] The partial evidence of the book of words potentially answers to the type of historical 'fragment' addressed by Gallagher and Greenblatt;[4] it was a subsidiary element of the performance text, and yet, as it is often the sole remaining version of the script, it moves necessarily to centre stage as a research tool. Certainly the sheer quantity of scripted references in the book of words provides a clear indication of the subject matter and the consistency of thematic focus addressed in any one pantomime. Even allowing for changes and omissions in performance, such a focus suggests strongly that many references, even after changes, remained in the performed version and influenced later additions. Researching provincial pantomime therefore requires not simply a historicist methodology of locating the scripts and reviews within their historical context, but also reading those scripts over a consecutive historical time period. Such a reading enables long term trends to be observed, and establishes themes and issues that were not only relevant for the contemporaneous audiences but were an essential component in establishing a recognisable identity for local pantomime, one that reflected the town or city in which it took place and which was carefully constructed by theatre managements to form a central part of the appeal and expectation of their potential pantomime audiences. Supporting evidence for the references is provided in the previews and reviews of the local newspapers which offer very relevant information, for example detailing visual images not noted in the book of words. Furthermore, with regard to political references the partiality of a reviewer guided by editorial policy can be usefully harnessed in a study of regional pantomime. Engaging with the intrinsic nature of the mediating reviewer and reading political preference into the reviews, enables voices of dissent and approval to further establish the local identity of the pantomime and the theatre.

During the second half of the nineteenth century, the increased number of pointed references, delivered by scripted, sung and visual means in performance, had become highly indicative of the new civic status of many of the growing towns of the period, reflecting their assertive industrial pride and political status. Foremost amongst those towns was Birmingham. Known popularly as the 'Workshop of the World', Birmingham's varied industrial output made it one of the most important British towns of the nineteenth century (it was given city status in 1889), and many considered it as second only in importance to London. The pantomimes produced at the Theatre Royal in Birmingham reflected and celebrated that economic status, but from the early 1870s the focus of topical comment also began to emphasise the civic and political status of the town that had been brought about by the work and influence of Joseph Chamberlain.

Chamberlain was not a local man, but after settling in Birmingham in the 1850s he had established himself as a successful businessman who was also interested in social reform. In 1869 he co-founded the National Education League, the motivating force behind the Education Act of 1870, which transformed the provision of primary education, and in 1873 he was elected Mayor of Birmingham, a post he retained for three successive years.[5] As Mayor he instigated and oversaw major improvements and redevelopment in the town and his stance as a Radical Liberal symbolised the political developments that were to dominate Birmingham for the remainder of the nineteenth century. In 1876 Chamberlain became a Member of Parliament for Birmingham and although he now moved on the national political stage, he retained close links with Birmingham and its civic developments.[6] That allegiance was reciprocated by the town; Chamberlain's status and that of the town were closely linked and his progress was charted with unfailing interest by the principal theatre, the Theatre Royal.

On gaining the civic seat in 1873, his achievement and establishment in the town was celebrated in the book of words for that year's pantomime *Beauty and the Beast*. A fictional cast list was headed by 'A Chamberlain of great renown' to be played by 'Mr. MAYOR'. The blurred boundary between the world of pantomime and the reality of nineteenth-century civic politics was a recurring feature of the productions at the theatre, notably in versions of *Dick Whittington*, to which I will return later in the chapter. However, the pantomimes also contained direct political allusion and comment. The 1873 production did not simply celebrate Chamberlain's new Mayoral role; in the opening dark scene, an exchange between the Spirit of Discord and the Sphynx extended the theme:

DIS. A *mayor's* nest you've disturbed, but to our gain
He has too long in Council *Chamber lain*,
A nobler destiny awaits him. He
May be the Sheffield blade's keen edged M.P.
He's full of work, and will some day be –

SPHYNX. Hold!
That future may be guessed at, but not *told*.[7]

This speech not only celebrated Chamberlain's immediate success but referred to the fact that he was due to stand for the Sheffield parliamentary seat in the forthcoming 1874 general election. The lines further engaged the audience in thought-provoking word-play. The phrase 'He/May be' asserted their role as electorate: Chamberlain's future would indeed be up to them, but the next line, 'will some day be –' was suggestively completed in performance by visual representations of Chamberlain together with leading figures from the current Liberal Government and Cabinet including John Bright, the long serving Birmingham MP, Prime Minister Gladstone, and Lowe, the Chancellor of the Exchequer. Disraeli, the leader of the Conservative Opposition also made an appearance, but the visual emphasis was on Liberal party power. Therefore the combination of political characters and the open-ended speech directed the audience to consider Chamberlain's longer term and 'nobler destiny'. The scene clearly indicated that Chamberlain *would* take his place in the political pantheon in the future, even if the specifics could only be 'guessed at, but not *told*'. According to strictures laid down by the Lord Chamberlain, visual representations of public figures were banned from the stage, a requirement frequently ignored by the Birmingham theatre. No complaint was made to the local theatre licensing authorities or to the Lord Chamberlain's office, and the only voice of dissent was that of the Conservative *Gazette*, which, in its review of the opening night performance, took the opportunity to

> enter our protest against the political element in pantomimes [...] let us have our annual treat with our little ones without being called upon to applaud a bid for any political party. Here we have Bright eulogised – Chamberlain idolised and Dizzy told he is a humbug. Speaking not as a Conservative, but as a pantomime-goer, we protest against it. Heaven knows we have enough of politics in most matters.[8]

It is unlikely that this particular reviewer was speaking purely as a pantomime-goer, but other members of the audience seemed undisturbed by the references and the correspondence columns of both the Conservative and the Liberal papers remained empty of accusations or counter-accusations regarding the pantomimes throughout the period. Chamberlain was in fact defeated at the 1874 election but idolising him was, it seems, *de rigueur* at the Theatre Royal. Birmingham was, historically, a Liberal town, and the theatre reflected that political stance. Indeed, references in the pantomimes did not flinch from following Chamberlain's adherence to Radical Liberalism.

At a formal dinner in October 1874, Chamberlain, as Mayor, had addressed members of the Birmingham Town Council. The Mayor and Council were only too aware of a forthcoming visit to the town by the Prince and Princess of Wales and Chamberlain followed established protocol at the dinner by proposing a toast to 'The Queen and Royal Family'. He then delivered a speech in which he declared himself loyal to the Royal Family and unwavering in his support of the proposed visit: 'a man might be a gentleman as well as a Republican; and even an advanced Liberal might not be unmindful of the duties of hospitality and the courtesy which every one owed to a guest'. He continued by defining his view of republicanism as that which was based on the 'unswerving faith in the value of representative institutions', decrying what was sometimes inferred in republicanism as the 'uproot[ing]' of 'existing orders'. However, his speech went on to denounce what he described as 'exaggerated loyalty' by those who 'enshrined [...] royal personages in [...] a stifling atmosphere of fulsome adulation', and who 'grew hysterical at public banquets on the occasion of the usual loyal toasts'. He also stressed his belief in meritocracy, and his speech appealed for public praise to be directed to the royal family 'for what they saw in them worthy of honest admiration' rather than 'attributing to them immaculate perfection and superhuman virtues'.[9]

Chamberlain's reputation and radical ideas concerned his critics prior to the visit and it was with noticeable relief that *The Times* was able to report on the 'heartiness and loyalty' with which the royal party were greeted when they visited Birmingham on 3 November.[10]

The royal visit and Chamberlain's republicanism were made the focus of an entire scene in the pantomime of that year, *Ride a Cock Horse to Banbury Cross*. In reality, the Prince and Princess of Wales had been met at the Birmingham borough boundary by the Mayor and Mayoress, together with a full reception committee and two troops of the 12th Lancers, prior to a procession through the town. In the pantomime, the retinue of the fictional Earl of Warwick was greeted by the character of the 'Mayor', who was unmistakeably costumed to represent Chamberlain. During the ensuing dialogue the former asked:

> And now – the question – I must plainly ask –
> Is yours a pleasant duty or a task?
> 'Tis said that you and others of your town
> Would stamp out monarchy, and put kings down
> If that be true our visit might perplex you.[11]

The scene here echoed both the contemporaneous occasion and contextual views of the local and national press and yet the pantomime script went further. The Mayor's response to this question – much as Chamberlain's initial speech to the town council – was not limited to a simple declaration of loyalty:

> We hope you will not let such rumours vex you.
> The sweets of office, and a Prince's smile
> Have made us very loyal – for a while.
> And now I say, though some around may frown,
> I hope we always may possess a *crown!*[12]

Whilst the exchange was distanced by its setting within the nursery tale and historical location of medieval Warwickshire, the references were unmistakable. By including this scene in the pantomime the theatre management had forced Chamberlain's views to the fore, refusing to distill his radical version of Liberalism either to assuage concern or in the name of amusement. The daring temporariness of the phrase 'for a while' accentuated Chamberlain's comments regarding the 'duties of hospitality', and the final line – 'I hope we always may possess a *crown!*' – concisely reiterated the closing remarks of his speech to the council in November in which he had supported the idea of monarchy, but concurrently inferred that that individual monarchs should prove themselves worthy of that place. This verification of Chamberlain's political views failed to raise any eyebrows in the press reviews, merely amusing the critics of all three of the main Birmingham newspapers. The lack of comment acknowledged the perceptive nature of the reference, but also signified an awareness of the theatre's political allegiance.

That allegiance to Chamberlain's Radical Liberalism recurred in pantomimes throughout the period. For example, in the 1881 pantomime of *Beauty and the Beast*, a pasteboard model of a 'huge lion' interrupted the proceedings of one scene. This moment is not evident in the script but was described in reviews: one character fought and killed the 'lion', and after being successfully destroyed and cut in half, 'the severed portions then turn[ed] round to the audience and display[ed] – one a portrait of Bright and the other one of Chamberlain'.[13] The representation of a lion connoted the heraldry of England, and, in showing pictures of John Bright and Chamberlain inside the destroyed lion, this image offered a layered interpretation. At one level it foregrounded the two Liberal MPs as political representatives of Birmingham, a town in the centre of England and at the centre of national industrial production, but it could also be read as a purely political image, of Liberalism as the core of English politics. The Liberal party had been returned to power in the general election of 1880 and that year had seen Chamberlain enter the Cabinet – the centre of British Government – for the first time. A third reading of the image might also have suggested that the destruction of the lion symbolised the defeat of the Conservative party, historically recognised as the representative party of the English Establishment.

Such was the fervour of support for Chamberlain by the management at the Theatre Royal that even errors of judgement were ignored. In the 1879 pantomime of *The Fair One with the Golden Locks*, a whole scene was dedicated to the industrial success of the town and the social improvements instigated by Chamberlain. His financial skill was praised by the character 'Brum', who stated: 'Thanks to our *Chamberlain's* financial skill/We have just now an overflowing till.'[14] Despite the practical achievements of the town improvements, including the redevelopment of slum areas, the expenditure had in fact straitened the town's municipal finances, a point of interpretation not missed by the *Gazette*, whose reviewer read 'pointed sarcasm' into the speech.[15] The praise in the pantomime was understandably underscored by the Liberal papers which acted to reinforce Chamberlain's achievements rather than focus on the financial difficulties, which were in effect denied in the pantomime.

The town improvements in Birmingham had earlier been addressed in the pantomime of 1875, in a production that provided early indications that the theatre was transforming Chamberlain's career into a local legend. In the opening scene of *Puss in Boots*, a lengthy speech outlined the work done in the town, concluding that 'For this advancement, thank brave Chamberlain/The local Whittington – three times Mayor – / Good luck to him. May he be long "all there".'[16] Chamberlain's three terms as Mayor of Birmingham were seized on by the theatre management as the rationale to stage regular productions of *Dick Whittington and His Cat*, the legend of the lad who travelled to London to seek his fortune and became Mayor of London three times. Through specific references the productions celebrated the very real achievements of Chamberlain, but by drawing attention to the Birmingham Civic Chair they also managed to assert the provincial town over London. Although in the staged pantomime stories, Dick Whittington indeed became Mayor of London, the emphasis remained on the Birmingham Mayoral office. In the 1884 production, that focus was emphasised by Whittington's arrival as Mayor of London being accompanied by a spectacular procession of Birmingham and Midlands trades.

It was upon Chamberlain's entering the Cabinet as President of the Board of Trade in 1880 that the pantomime of *Dick Whittington* was performed for the first of three occasions in the 1880s, all making a very specific connection between the hero of the legend and the hero of Birmingham:

> Lord Mayor *three* times! And may it be my fate,
> To be an M. P. and Minister of State!
> The thought of that makes me so gay and *bright*
> I might have in a *chamber lain* all night![17]

The theatre management engaged the successful music-hall star Vesta Tilley to play Dick in both the 1884 and 1888 versions of this pantomime. In so doing, her repeated appearance in the role established a sense of continuity and a reiteration of sentiments that underscored Chamberlain's own successive achievements and local popularity. In the 1888 version of *Dick Whittington*, the character of Dick in the final triumphant scene appeared on a real white horse; the civic hero defeating wrong, surviving slander and assault in the world of the pantomime at a time in the late 1880s when Chamberlain, Birmingham's own civic legend, was fighting political battles in Westminster; in 1886 he had left the Liberal party over the issue of Irish Home Rule, and aligned himself with rebel Conservatives to form the Liberal Unionist party.

Chamberlain's dominance of civic matters in Birmingham engendered a powerful sense of allegiance in the town that was eagerly reflected in productions at its oldest established theatre. Such an affiliation may have been aligned to personality as well as the practical effects of his social reform, but the admiration of Chamberlain in the pantomimes was not simply the effect of a single personality. His alliance with Conservative MPs after 1886 did not affect the regard with which he was held in Birmingham, but nor did it alter the political stance of the Theatre Royal pantomimes. Liberal political support ran deep in Birmingham and although the Chamberlain years of the 1870s had a ready-made hero for its pantomimes, the theatre continued to express political support for the party throughout the 1880s, which can be evidenced in periods of Conservative as well as Liberal Government. Although pantomimes have traditionally enjoyed jokes at the expense of a range of public figures and political leaders, the Birmingham pantomimes frequently contained negative comments about local and national Conservative figures that were not always confined to humorous or harmless jokes. For example, in the aforementioned 1874 pantomime of *Ride a Cock Horse to Banbury Cross,* the character of 'Guy, Earl of Warwick' delivered a long speech in Scene 7, in which he promotes political hypocrisy:

> To 'pubs' make promises; abstainers laud;
> The church uphold, and then Dissent applaud. [. . .]
> Go in for working men; for masters stick up,
> And o'er every grievance make a 'kick up'

The speech could have passed as a general and satirical commentary on election promises, but it was concluded by the character of the 'Mayor' – already established as representing Chamberlain – stating of Sir Guy, 'From those strong sentiments, thus indicated,/It strikes me, you are growing *Dizzy – pated!'* [18] The implication of the speech was therefore instantly shifted from one of general cynicism to a specific attack on the policies of 'Dizzy', or Benjamin Disraeli, the Conservative Prime Minister. The following year, after the lengthy speech praising Chamberlain's town improvements, characters concluded that 'Our Tory rulers have not brightly shone'.[19] And the 1876 pantomime of *Sinbad the Sailor,* with its eastern setting, provided ample opportunity to criticise Disraeli for his foreign policy in the Balkans. A more local example could be found in the Conservative Alderman and former Mayor, Henry Hawkes, who was an active opponent of Chamberlain. Hawkes was treated with disdain in the local Liberal press and became a regular target in the pantomimes. For example, in the 1882 pantomime of *Sinbad the Sailor*, the demon character of the Old Man of the Sea cursed the Liberal Caucus – the ruling civic body of the local Liberal party that had been established by Chamberlain – 'But I will kill it, since my plans it baulks,/And leave its *Car-cus* to be pecked by *Hawkes*.'[20]

While responses to the pantomime references by local reviewers were naturally aligned to the political stance of their newspapers, complaints by those reviewers about comments and images in the pantomimes were on the whole limited. It might be assumed that regular attacks on the Conservative party leaders and policies and local civic dignitaries would have resulted in more regular responses from the *Gazette*, but in general the paper refrained from engaging with what may have been judged adverse comments. Such an attitude – as I have already suggested – implies that the newspapers were perfectly aware of the theatre's stance on political issues; it was well known for example that the theatre manager as well as two of the principal writers of the Birmingham pantomimes in this period – Charles Millward and James J. Blood – were themselves Liberal supporters. Millward aligned himself to Radical Liberalism and Blood was a member of the Liberal Arts Club in Birmingham, of which Chamberlain was for a time Chairman. The Arts Club had been 'initiated under the auspices of Mr. Mercer Simpson' the manager of the Theatre Royal, and its members made use of rooms on the theatre premises.[21] An acceptance of the political preferences in pantomime topicality at the Theatre Royal may therefore explain the relatively subdued response from the Conservative *Gazette*. The local Liberal papers were naturally in accord with the sentiments expressed in the annual pantomimes, in particular the *Birmingham Daily Post*, whose owner, J. T. Bunce was a noted local supporter of Chamberlain. However, an inclusion to the pantomime of 1887 prompted an unusually negative response in the *Post*. That response further highlights both the expectation of political affiliation and the manner in which the pantomimes staged that affiliation.

In December 1887 the theatre management staged the pantomime of *Little Goody Two Shoes*, starring Miss Jenny Hill, the 'Vital Spark' of the music halls. In the newspaper reviews of the opening night performance, Hill's patriotic song 'The Old Flag' was applauded by all the Birmingham papers, but her topical song 'England for the English' was less well received by both the principal Liberal papers. The *Birmingham Daily Mail* thought it contained 'bunkum words and tawdry sentiments', thankfully glossed over by Hill's performance of 'dash and spirit', but the *Birmingham Daily Post* was incensed:

> [as] for a song in praise of the many virtues of Fair Trade – well, we suppose that it is hopeless to expect that music-hall performers will ever understand how ridiculous is the figure which they cut in their attempt to pose as the political instructors of their audience. However, we have no doubt that Miss Hill knows as much of political economy as the majority of the Fair Trade party.[22]

The *Birmingham Daily Gazette* however applauded the 'patriotic song, in which Free Trade, "all one-sided though," comes in for a very drastic denunciation'.[23] Why Miss Hill's song divided the critics depended on that subtle difference between the words 'Free' and 'Fair' Trade. Free Trade was a policy that had been adopted by both political parties in the course of the nineteenth century, but by the mid-1880s there was a more defined division between the Liberal adherence to Free Trade, and the protectionist measures – or Fair Trade – which were championed by members of the Conservative party. During the national depression of the 1880s, the varied trades and industries in Birmingham were not universally affected, and whilst Free Trade did encourage foreign competition, Liberal opponents of Fair Trade were more concerned about the rising prices of necessities, especially food, which might have occurred if protectionist trade tariffs were introduced. There had been considerable debate in Birmingham in 1887 regarding the two policies, the Free Trade argument being led by letters in the press from the longstanding Liberal MP John Bright. The divided reception of the pantomime song by the reviewers was therefore naturally aligned to the political stance of their newspapers, but the extended attack in the *Post* was also a complaint against Conservative principles being introduced in relation to heroic sentiments. The Theatre Royal pantomimes contained a variety of political references from overt caricature to subtle asides, but a singular trend in this period was to align party politics to character types. In other words, the heroes supported Liberal policies and politicians whilst the human villains and supernatural demons claimed Conservative principles. This dichotomy was presumably intended in the 1887 pantomime: the villain was named King Counterfeit, a Conservative who, when defeated in battle at the end of the story, remained adamant that he '*[s]tuck to my colours, and so dyed true blue'*.[24] He and his cohorts attacked Liberal policies and at one point he disguised himself as a pedlar, selling items representing the Liberal party, such as 'Gladstone axes' and Chamberlain monocles, a reductive symbolism which suggested that Liberalism comprised only token power and political representation. Conversely, the Principal Boy, 'Sterling' represented the plucky and hardworking Liberal Birmingham working man. The heroic Liberal character type had been effectively represented by Vesta Tilley in the 1879 pantomime, *The Fair One with the Golden Locks*, and again in the two productions of *Dick Whittington* in 1884 and 1888. Therefore, although an anti-Free Trade song could have been sung in the 1887 production, it should not have been sung by the hero. The song, copyrighted to Hopwood and Crew of London, was not written by the author of the pantomime James J. Blood, and represented an uneasy inclusion that clearly failed to take account of the complexity of the trade issues that had been debated in Birmingham during the preceding months. The complaint voiced in the *Post* review did not criticise the theatre management or the author, but steadfastly focused on the political illiteracy of the 'music-hall performer'. In particular the reviewer attacked Hill for what he perceived as her 'attempt to pose as the political instructo[r]' of the audience, a statement that raises an important issue regarding the reception of political references by audiences at the Theatre Royal.

Pantomime audiences have long been recognised as reflecting a large cross-section of society, appealing to all classes and ages. Further, as Susan Bennett has indicated, each member of a theatre audience would have experienced the production according to their horizons of expectation, some very individual and some shared.[25] In terms of audience reaction to political references, especially the extended comments at Birmingham, different responses were occasionally noted by critics. For example, in the 1868 pantomime of *Ali Baba and the Forty Thieves*, an illustration of 'leading personages', including Disraeli, Gladstone and Bright, elicited both 'cheers and counter-cheers' from the audience.[26] As a major theatre in a large town, attracting audiences from a wide topography across city, suburbs and county as well as further afield, the Theatre Royal management could not dictate the political nature of those audiences and naturally a range of political understanding would be part of a large audience's cultural experience. However, even if individual members of the audience held varying political ideas, the sustained political emphasis of the Birmingham pantomimes would have become part of the range of expectations experienced by many of the theatre-goers visiting the pantomime. A large section of the regular theatre-goers at the Theatre Royal would have begun, over time, to expect a Liberal emphasis in the political referencing of the pantomimes. The correspondence columns of the local newspapers charted lively debates from a range of readers, from politicians to urban working men and farmers, but references in the local pantomime prompted no written responses. Despite the fact that topical speeches were reprinted in previews and reviews to alert readers to political referencing and to provide suitable interpretations, potential and actual audience members remained apparently unperturbed by the attitudes expressed in the Christmas production. Similarly, no local complaints were made to the Lord Chamberlain's office about any of the Theatre Royal pantomimes in this period, all of which perhaps suggests a level of complacency by Birmingham audiences. Both David Mayer and, more recently, Millie Taylor have noted how pantomime has always 'poked fun at authority figures'[27] but the references at Birmingham could range from quite serious accusations to simple caricature. If those references had engaged in simply 'pok[ing] fun' at all political parties and politicians, the occasional more pointed attack might have caused comment from members of the audience, but the sustained tone combined with the lack of complaint suggests strongly that, even allowing for a range of political support amongst the audiences (both 'cheers and counter cheers'), there existed an established knowledge by many that that theatre supported the Liberal party. Such knowledge made the audiences not complacent, but complicit. Within that context of expectation and complicity, the accusation made in the *Birmingham Daily Post* that Jenny Hill's anti-Free Trade song was an attempt to educate the audience, takes on a different tone. If many in the audiences at the Theatre Royal were aware of the political motivations behind the pantomime referencing, the fear expressed by the *Post* was perhaps, that Hill was trying to *re*-educate them, against the established principles which governed the remainder of the production.

At the start of this chapter I noted the similarities between the expectations of nineteenth- and twenty-first century pantomime audiences, but the issue of topicality in terms of coverage, style and tone becomes much more complex when discussing nineteenth-century provincial productions. Millie Taylor, in *British Pantomime Performance*, briefly addresses the political topicality in modern productions which, she emphasises, tends to short references presented in a 'superficial and humorous way'.[28] In contrast, the political comments in the pantomimes of Victorian Birmingham were of a much more detailed and sustained nature. In particular, the pantomimes of the 1870s and 1880s contained a succession of political references in a variety of forms, from caricature to characterisations, from lengthy debates and direct comments to the layered encoding of visual imagery. Those references charted a partisan approach to political developments at a national level but also, and more importantly, were able in the figure of Joseph Chamberlain and his version of Radical Liberalism to underscore the industrial and political status of Birmingham. Although the political views expressed in the pantomimes were at times unsubtle and outspoken, the productions worked within a complicit environment in their reflection of local knowledge, events and power struggles.

Notes

1. The principal work in this field remains David Mayer III, *Harlequin in His Element: The English Pantomime 1806–1836* (Cambridge, MA: Harvard University Press, 1969). See also Jane Moody, *Illegitimate Theatre in London 1770–1840* (Cambridge: CUP, 2000), Chapter 7, 'Illegitimate Celebrities', pp. 208–341.
2. See for example: Peter Holland 'The Play of Eros: Paradoxes of Gender in English Pantomime', *New Theatre Quarterly*, 13.51 (1997), 195–204 and Jim Davis 'Imperial Transgressions: The Ideology of Drury Lane Pantomime in the Late Nineteenth Century', *New Theatre Quarterly*, 12.46 (1996), 147–155.
3. For a more detailed discussion of the Victorian book of words and its role in audience behaviour, see Jill A. Sullivan, 'Victorian Pantomime Libretti and the Reading Audience' in Graham Allen, Siobhán Collins, Carrie Griffin and Mary O'Connell (eds), *Making an Audience: Reading Textual Materiality* (forthcoming, Pickering & Chatto, 2010).
4. Catherine Gallagher and Stephen Greenblatt, *Practicing New Historicism* (Chicago and London: University of Chicago Press, 2000), p. 19.
5. H. C. G. Matthews and Brian Harrison (eds), *Oxford Dictionary of National Biography* (Oxford: OUP, 2004), vol. 10, pp. 924–5.
6. Ibid.
7. *Beauty and the Beast* (1873), book of words, Scene 1, p. 3. Accession no. L28.1 64589, Birmingham Local Studies Library.
8. Theatre Royal', *Birmingham Daily Gazette*, 27 December 1873, p. 5.
9. 'Dinner to the Birmingham Town Council', *Birmingham Daily Post*, 7 October 1874, p. 5.
10. Quoted in 'the London Press on the Royal Visit', *Birmingham Daily Post*, 5 November 1874, p. 6.
11. *Ride a Cock Horse to Banbury Cross* (1874), book of words, Scene 3, p. 9. Accession no. L28.1 314833, Birmingham Local Studies Library.
12. Ibid.

13. 'The Theatre Royal Pantomime', *Birmingham Daily Post,* 27 December 1881, p. 5.
14. *The Fair One With the Golden Locks* (1879), book of words, Scene 1, p. 5. Accession no. L28.1 64592, Birmingham Local Studies Library.
15. 'The Theatre Royal Pantomime', *Birmingham Daily Gazette,* 24 December 1879, p. 5.
16. *Puss in Boots* (1875), book of words, Scene 1, p. 3. Accession no. L28.1 64591, Birmingham Local Studies Library.
17. *Dick Whittington* (1881), book of words, Scene 4, p. 12. Accession no. L28.1 64593, Birmingham Local Studies Library. '*bright*' in line 3 alludes to John Bright, who was a Liberal MP for Birmingham from 1857 until his death in 1890. He was often linked with Chamberlain in pantomime references.
18. *Ride a Cock Horse to Banbury Cross* (1874), book of words, Scene 7, p. 16.
19. *Puss in Boots* (1875), book of words, Scene 1, p. 4.
20. *Sinbad the Sailor* (1882), book of words, Scene 1, p. 2. Accession no. L28.1 243340. Birmingham Local Studies Library.
21. E. Lawrence Levy, *Birmingham Theatrical Reminiscences: Jubilee Recollections (1870–1920)* (Birmingham: Messrs. J. G. Hammond & Co. Ltd. [1920]), pp. 29, 89.
22. 'The Pantomimes: *Goody Two Shoes* at the Theatre Royal', *Birmingham Daily Mail,* 27 December 1887, p. 4, and 'Theatre Royal – *Goody Two Shoes*', *Birmingham Daily Post,* 27 December 1887, p. 5.
23. 'Theatre Royal. *Goody Two Shoes*', *Birmingham Daily Gazette,* 27 December 1887, p. 5.
24. *Little Goody Two Shoes* (1887), book of words, Scene 7, p. 27. Accession no. L28.1 125961, Birmingham Local Studies Library.
25. Susan Bennett, *Theatre Audiences: A Theory of Production and Reception* (London: Routledge, 1997), pp. 125–65.
26. 'Christmas Amusements: Theatre Royal, *Ali Baba and the Forty Thieves*', *Birmingham Journal,* 2 January, 1869, p. 5.
27. Millie Taylor, *British Pantomime Performance* (Bristol: Intellect Books, 2007), p. 142.
28. Ibid. p. 137.

10

'Holding up the mirror': Readership and Authorship in the *Era's* Pantomime Reviews from the 1870s

Ann Featherstone

From Covent Garden or Colchester or Coatbridge, if a performer or manager wanted to review the state of the theatrical world in the 1870s, she or he would invariably turn to the *Era*.[1] Indeed, as the correspondent claimed in his column 'The Christmas Novelties' for 1878:

> if any future historian should want to know a hundred years hence what was the social life of England at this period of the century, he could scarcely avail himself of a more comprehensive descriptive guide than he will find in the pages of the week's *Era*.[2]

It was the foremost trade journal for the profession, in which the week's theatrical affairs were noticed, business was discussed, engagements detailed and where anything from stage beards to potions to relieve distemper in stage-dogs was advertised. From its front page (which, by the 1870s, bore the 'cards' of the legitimate acting profession), to its columns on the American and Parisian stages and those reporting on the provincial theatres, reports on the music halls and musical concerts, and its editorials, considering the profession and the 'theatrical times', it was the mirror held up to reflect the business of entertainment. Concern – about the state of the profession, conditions of employment, the direction which performance of all varieties was taking – taxed its contributors, and throughout the 1870s this included the legal wranglings between theatre and music hall management, performers and writers, over what constituted a stage play, and the voicing of worries about the adverse influence of French plays upon the impoverished home-grown output.

The *Era*'s first issue, published on 30 September 1837, was not in the least inclined towards the theatre. It was a Sunday paper, with the front page devoted to foreign matters and the remainder containing accounts of 'prize fights, race meetings, sport and politics, with just a dash of drama'.[3] Over the next 70 years, the Ledgers, Frederick and his son Edward, steered the paper towards the comprehensively dramatic, and after his father's death in 1874, Edward Ledger (Figure 10.1) discontinued the columns reporting sporting events, the last vestiges of the early *Era*. It was a journal noted as a 'cradle of the critics', a training ground for aspirant young journalists. Clement Scott did some of his earliest work for the *Era*, recalling his excitement at the delivery of the ticket in a blue envelope for anything from an amateur performance, to a concert or reading.[4] H. G. Hibbert, perhaps best known for his book *Fifty Years of a Londoner's Life* (1916), was the *Era's* Birmingham correspondent and E. L. Blanchard, author of 36 consecutive Drury Lane pantomimes, wrote consistently for the paper from 1850 until 1889. H. Chance Newton, A. B. Walkley and J. T. Grein, all eminent authors, journalists and critics, similarly contributed over the decades when the *Era* became

Figure 10.1 Mr Edward Ledger, *Judy*, 21 January 1880 (collection of Ann Featherstone)

affectionately known as the 'Actor's Bible', the 'Organ of the Profession', and Charles Dickens's 'favourite journal', the delivery of which ('dropped down into his area') he eagerly anticipated each Sunday morning.[5] It continued as a Sunday newspaper until 1880, when publication day moved to Saturday, and it finally ceased publication in 1939.

Dickens claimed that he derived his 'extraordinary stock of theatrical information' from a close perusal of the *Era*, and certainly it was essential for anyone desiring a comprehensive understanding of the stage in all its forms.[6] (Clement Scott claimed that any writer attached to the *Era* had to be 'conversant with the whole history of the stage, from the mystery plays to the days of Dion Boucicault!'[7]) But, as Dickens illustrates through a fantastical creation based upon the *Era*'s Classified Advertisements columns, only an adept could hope to negotiate the esoteric language with which the stage spoke to itself. For anyone else, 'worse than Greek, Hebrew, and Double-Dutch', says Dickens 'would be these advertisements to you'.[8] Even Manchester's satirical weekly, *The Sphinx,* commented upon the insularity of the journal – and the theatrical profession in general. 'To anyone taking up the *Era* for the first time', it says, 'its contents are a new revelation of life; they are a glimpse into a busy, teeming existence which is, and has been, going on close by his side, unnoticed and unknown.'[9] Dickens, of course, claimed he was 'cradled in a property washing-basket [and] nursed by a clown', and thus a fluent speaker of theatrical argot, and regarded himself as a graduate of 'the university of the great theatrical newspaper'.[10] H. Chance Newton declared that he learned to read by the *Era*.[11]

Both Dickens and the anonymous writer in *The Sphinx* beg the question, then, of who constituted the readership of the *Era* and, in particular, who contemplated those reports of provincial and metropolitan pantomime which occupied so many column inches for so many months. Were these highly particularised reviews targeted at a readership of professionals, the Bluff Kings, Saucy Knaves and Princess Cherry-ripes to whom Frederick Ledger wished 'A Merry Christmas' and who, by the last week in January, would be anxiously scanning the 'Wanted' columns for their next engagement?[12] Or were they essays, self-conscious advertisements for the glories of Drury Lane or the peculiarities of the Britannia, Hoxton, born out of insider loyalty to stalwarts of the profession? And what of the amateur and provincial offerings, many of which warrant only a scant notice or, worse, a scathing one? Who did the columnists imagine was reading their reports? Professionals? Amateurs? Managers? Audiences who needed persuading? or those who wondered if the writer had seen what they had seen?

One of the most interesting features of the *Era* is its catholicism – from the lowliest portable theatre to Drury Lane magnificence, it embraces all. Neither Mr Colville's pantomime of *Jack and Jill* written for his Pavilion portable in Mansfield, Nottinghamshire nor Mr Blanchard's twenty-fifth 'annual' for Drury Lane (*Aladdin; or, Harlequin and the Wonderful Lamp*) escape the notice of the *Era's* reviewers in 1874, but that is not to say that both are treated in the same way, in equal detail or with equal generosity. The *Era* published, even in its early days, provincial as well as London reports of what it termed 'entertainments', but interest was, unsurprisingly, focused upon the metropolis, widely regarded as the 'common centre of the profession'.[13] Although the Preface to the report on the Christmas Novelties in December 1879 noted that 'the Provinces now compete with the capital in brilliancy of theatrical decoration, and at several of our larger towns and cities, the Christmas pieces are even more elaborately embellished than those of the metropolis', the column inches devoted to the London pantomimes far exceeded those allocated to provincial accounts.[14]

Cities and towns outside the capital clung to their pantomime traditions. After the demise of the circuit companies, and when the old eighteenth-century theatres fell into disrepair or were unavailable, the pantomime was forced to find an alternative venue, a town hall or assembly room. Derby had its Lecture Hall and Corn Exchange, Barnsley its Circus of Varieties, but neither venue nor company could compete with a Drury Lane or an Adelphi show. Even the newly-built theatres – Manchester's Prince's Theatre (1864), for example, or Liverpool's Rotunda (1860) – with their up-to-date facilities, could not compete, for the *Era* correspondents, with the reputation, fictive and real, of London. Reporting on provincial pantomime was particularly hampered by the limited wordage allowed, compounded by the miniscule font (often only 7 or 8 point) in which their finished pieces were reproduced. Given these restrictions, reviewers were forced to choose between a simple list of the *dramatis personae* from a handbill, or to attempt a succinct, often abbreviated account of the pantomime and performance. Whilst listing names and roles located the professional in a venue and company to the wider professional community, this was the provincial correspondent's *least* reflective response to pantomime. As the run continued – and eight weeks was the maximum in the provinces – the report might extend to identifying leading players, and contain a range of standard phrases – 'the most gorgeous', 'the most splendid', 'exceeding expectations'. But 'This is the best Pantomime we have had here for many years' was a frequent resort.

The dullness of most provincial reports from the 1870s suggests that the intention was merely to notice the performance, and that it served as an unpaid advertisement to the profession, of the theatre, company and individual artists, for whom *any* notice, however mild, was better than none.[15] As the numbers of provincial notices increased throughout the 1860s and 1870s, and particularly when Frederick Ledger focused the *Era* directly upon the theatrical (see above), this tendency to 'merely notice' performances was increasingly expedient, and really communicates little but the bare detail. So that when a critical point, positive or unfavourable, is made, it is instantly apparent, as in the case of the Northampton correspondent, who in two sentences, noted the Theatre Royal's pantomime of *The Fair One With the Gold Locks*, adding (with caustic conciseness), 'the role of the "fair one" being undertaken by a lady whose ringlets were as black as the raven'.[16]

Allusions to and jokes about local matters were vital to the regional pantomime and, for critics, were one of the defining differences between London productions and the regions. Whilst the cosmopolitanism of London demanded from its pantomimes '"business", and spectacle, and fun, and song and dance' which was reported in detail, the provinces were more introspective and parochial, audiences enjoying 'talk' at the expense of 'the authorities and the magnates'.[17] This preference is evident from the local reviews, and indeed, the 'want of localisms' (apparent in *The Babes in the Wood; or Harlequin Robin Hood and his Foresters Good, and the Brave Little Soldiers of Lilliput* at the Theatre Royal, Dewsbury) was regarded as a serious fault.[18] Plentiful 'local hits', refreshed at intervals, could make amends for an otherwise indifferent production. Whilst the local reviewer of *Sleeping Beauty; or, Jack and Jill and Harlequin Humpty Dumpty* at the Theatre Royal, Preston, was largely unimpressed by the 'wretchedly poor' dialogue, he felt it was redeemed by 'a few humorous local hits, which the audiences are quick to appreciate.'[19] The balance between spectacle (good scenery and costumes), dialogue (sharp, witty, inoffensive) and local material was the key to a successful show and a good notice. The combination of good scenery, dresses and 'appointments' in the Langley Moor pantomime, *Aladdin and His Wonderful Lamp* was enhanced by a dialogue which 'abound[ed] with comic allusions to local subjects', while in Nottingham Frank Green's *The King of the Peacocks* was a popular show once the performers, with practice, had become 'more perfect in their business' and with the 'topical songs' which contained allusions to local problems such as the adulteration of milk.[20] A judicious mix of the local and national was also a good formula. The *Forty Thieves* at the

Theatre Royal, Hanley was lifted by 'local and political illusions [*sic*], and an appropriate reference to the Northfleet catastrophe, and the sad fate of Captain Knowles'. Strangely, the burlesque of the Tichborne Claimaint was regarded by the Hanley correspondent as being 'out of place in a Pantomime'.[21]

More ambitious provincial dramatists took a successful pantomime 'book' and gave it a new lease of life. At the Theatre Royal, Blackburn, J. F. McArdle took H. J. Byron's *Lurline, the Nymph of Lurleyberg; or, Harlequin Rupert the Reckless, and the Spirit of the Dancing Waters* which had been performed at the Prince of Wales Theatre, Liverpool the previous year, and 'revised and localised' it. One of the most successful elements, according to the local correspondent, was a satire on the recent furore over the 'Can-can' which had appeared in Offenbach's opera-bouffe *Vert-Vert* at the St James's Theatre and caused the Lord Chamberlain to issue a circular to the managers of the London theatres, taking them to task over the 'debased state' of the English stage reflected in 'indecent dances and immodest dress'.[22] The Blackburn critic and the audience of the Theatre Royal clearly enjoyed Mr McArdle's satire:

> Six girls appear dressed in white from shoulders to feet, with night-caps on their head, and coloured spectacles on their noses, and, in the words of the text, go through a ballet strictly moral, much to the amusement of the audience.[23]

Many correspondents suggest particular improvements, evidence that at least they expected their comments to be read by managers and performers. The Manchester correspondent is very specific in his advice. The Skeleton Dance in *Aladdin the Great* was pronounced 'too grim for an entertainment certain to be largely visited by children' and in *Twinkle, Twinkle Little Star* Mr Edmonds was advised to:

> insist upon the elimination of the incidents which disfigure the Harlequinade, and two scenes in the Opening, viz. the Stretford Village scene and that of the picture gallery.

Whilst, in *Little Jack Horner,* he appreciates Mr Edgar's topical song and its 'epigrammatic hits at current events', he is firm that:

> the omission of an expletive which he introduces with a view to humorous effect is advisable in a house that is now patronised so largely by ladies in the best circles of Manchester society. [24]

This critical faculty and the freedom to express it seems to have been a *raison d'être* for some, like the Derby correspondent who bemoaned the fact that Mr Fairlie's production of *Cinderella* in the Lecture Hall was 'faultily faultless' and that he could 'only grumble at having nothing to grumble at'.[25] But he made up for it elsewhere. *Little Red Riding Hood* at the Corn Exchange was 'a miserable apology for a Pantomime', playing to 'empty galleries and deserted boxes'. 'We are sorry to be obliged to observe,' he says, 'that the merits of the production are on a par with those of a certain leg of mutton on which the illustrious Dr Johnson dined, while travelling from London to Oxford, and which he, with characteristic energy, pronounced to be "as bad as bad can be – ill fed, ill killed, ill kept and ill dressed"'.[26] Perhaps the harshest report in the 1870s comes from the reviewer of the Theatre Royal, Plymouth's pantomime, *Little Bo-Peep* which is:

> without exception, the very worst performed we have seen on our boards for many a season ... the acting and singing, particularly the latter, are truly wretched. Up to the time of writing, we have scarcely heard a note in tune, and when we say that the sheep are almost the best performers throughout, we only speak the truth.

At Oldham, the difficulties are with the 'book', and the critic notes, with some generosity, that '[the performers] do everything in their power to render effective, material which is not always of the highest order'.[27]

This well-tried technique of 'damning with faint praise' was a regular resort, but the delicate line between a poor production and one which, like all nineteenth-century pantomimes, was 'work in progress' had also to be negotiated by the reviewer. By the second week of January, it was expected that the machinery of a pantomime should be running smoothly and that any adjustments to the content should already have been made. Brighton Theatre Royal's *Hop o' my Thumb* is a good example of a show 'running in', the reviewer noting that 'since its first presentation there has been a judicious application of the pruning knife and a general smartening of the music, and everything now works smoothly and satisfactorily'.[28] But there are those productions which, even after a three or four week run, are clearly beyond redemption. In Portsmouth, the pantomime of *The Green Gnome; or, Harlequin the Prince of Utopia, the Pretty Princess, and Aurora, the Good Fairy of the Enchanted Wood,* given by Mr Hughes at his Palace of Varieties, had a luke-warm reception: 'The plot', says the reviewer, 'is a very misty affair, and hinges chiefly on the political events of the day. Occasionally the fun is spasmodic, but the author, who has written exceedingly well on all former occasions, has forgotten this time, the old adage, "A little nonsense now and then is relished by the wisest men".'[29] These relatively gentle criticisms are suggestive of shows which never quite realise the ambitious manager's aspiration: creaky mechanical effects, an uninspiring Transformation Scene, an impenetrable or unfunny 'book', the dullness of it all.

These provincial reviews describe, no doubt, perfectly adequate performances, and to simply 'notice' them was the brief of the *Era*'s correspondents. On the one hand they had free rein (within their word limit) to 'notice' whatever they liked – and there were numerous occasions when performers took them to task by letter, feeling that they had not been sufficiently noticed – but also they were self-policed, having a notional sense of 'good' performance and 'good' practice against which they measured what they had seen.

This notional sense may have been derived (rather unreasonably) from the London theatres and pantomimes, to which the *Era* devotes many column inches in intricate descriptions of, not only the convolutions of the pantomime story and the fourteen scene changes, but also the disposition of the audience and the vagaries of the London weather during the season. This is, indeed, another world, far removed from the trials and tribulations of sheep and misty plots. It is not to say, of course, that the metropolitan correspondents are uncritical of performers or performances but, unlike the provincial journalists, they rather tend towards mild observation – Jenny Hill's appearance as *Jack the Giant Killer* at the Gaiety is compared with Louise Keeley's and she is excused as 'nervous' though 'warming to her work'.[30] The Royal Victoria Palace (the 'Old Vic'), recovering from one of many overhauls in December 1871, was simply not ready for *The Bronze Horse*, and though the *Era's* correspondent felt that the performance should have been postponed, he was indulgent and prepared to excuse the many shortcomings. Even worse fortunes befell Mr Harcourt at the Royal Alfred, Marylebone, in the same year, and his *This is the House that Jack Built; or, Harlequin Pussy Cat, Where Have You Been? The Little Wee Dog and the Good Child's History of England*. Only part of the Transformation Scene could be shown; the masks had not arrived and the house scene 'did not work well'. Mr Harcourt, having apologised before the curtain three times, offered the public their money back but, the *Era*-man declares, 'Marylebone was still amiable, and accepted the Harlequinade without the Transformation'.[31]

This spirit of indulgence pervades much of the metropolitan reporting on pantomime, suggesting a different agenda and even addressing another readership. For if provincial performers and managers anxiously looked for a simple 'notice' mediated by forbearance, metropolitan professionals found responses to their performances embedded in a very different set of concerns. The evidence of the 1870s shows the fondness and generosity with which the leading newspaper of the profession regarded the London pantomime, and this pinnacle of the year (both critically and economically) is celebrated for its paradoxical adherence to tradition and progress. Of course, pantomime, whether reported in the *Era* or the literary journals, was never 'what it had been': the days of Grimaldi would not come again, 'the best features of past days … being driven from the modern stage', but the efforts of dramatists such as E. L. Blanchard at Drury Lane sustained at least those tender connections with childhood innocence and delight.[32] Progress, embodied in the technological wonders of stage machinery and effects, compensated for the passing of the simple pleasures of yesteryear. The child-become-man might experience again the heady delights of Boxing Night.

Not all was gilt and gaudy. The *Era's* editorial and its prefatory paragraphs on 'The Christmas Novelties' reported upon the weather (always a concern for London managers) and, particularly in the editorial, conducted a framing exercise in which the gaiety of the season is set against the difficulties and injustices of the real world. Poor people were starving, were 'shivering with cold or reeling with gin'.[33] In 1874, having played with the idea that the solemn and grumpy men of the City should and do 'give themselves up to innocent enjoyment' at the pantomime, the editorial issues a solemn reminder about the hard-working professionals who provide the entertainment and who 'have not been able to sit quietly down to a Christmas dinner surrounded by genial friends and relatives'.[34] This very attractive motif of reality set against fantasy is one which reviewers use frequently – but not as frequently as the pervasive discourse linking pantomime and innocence: the re-creation, the revival and the resurrection of childhood.

If there is a difference between the journalistic models of the *Era's* provincial and metropolitan pantomime audiences, then it lies in the presence, real or imagined, of children. Whereas the provincial reporter apparently has little awareness of juvenile spectators, the West End theatres are (imagined to be) populated by 'the little people home from school for the holidays'.[35] The embedded regret and longing for childhood and innocence appears in concentrated form within these pantomime reviews, and the editorials which precede them. For example, the editorial for Christmas 1871 invokes, once again, that audience of 'our boys … home for the holiday' – reclaiming them from their public schools to the family, the paterfamilias and Christmas. The editorial continues in mounting hyperbole invoking the anarchism of the Harlequinade in the excited voice of the man-child – 'Tell us, Mr Clown, where you got those supple legs of yours, not a bone in them, as it seems to us. Knees in or out, calves fore or aft, joints bending all ways; and back bone as elastic as Indian rubber,' concluding, 'Ah! let us never know how these things are, any more than the pain and labour of doing them; but let us live year by year in the dream of childhood, wonder-worshippers in front of the Pantomime scene.'[36] If this expression of wanting to blindly embrace the illusion of pantomime contrasts with the earlier reminder of the human cost in the labours of the 'hard-working professionals', it also hints at the pervasive fascination with demystification. The popular and literary journals describe in revealing detail the processes involved in 'getting up a pantomime' for the illumination of the non-professional adult reader, from the setting of Clown's traps and trick-boxes, to the revelation that Columbine is a mother with small children![37]

London pantomime, particularly in the West End, is located within a cultural frame in which class and gender are viewed at play: in the editorials, the child audience is invariably 'home for the holidays' and, therefore, like the journalist, male. Unlike the Drury Lane 'Stentors' in the gods, who throw orange peel and roar out the latest music hall song, conducted by one of their number in his shirt-sleeves, the little men in the circle are well-behaved. For the critic, the success of a pantomime is measured by its capacity to enchant child spectators, who are drawn with a vocabulary which, year on year, remains locked into unchanging childhood:

> Little heads bent eagerly forward, little eyes sparkling with delight, little throats giving forth peal upon peal of laughter, little fingers pointing, and little hands clapping.[38]

The Drury Lane pantomime is singled out for particular treatment in the *Era*. Nowhere else is pantomime written about with quite such undisguised effusion – for the form, the subject, and indeed the author, E. L. Blanchard (1820–1889), who by the 1870s had been writing 'Annuals' for Drury Lane for over twenty years. The eulogistic approach to both dramatist and performance is, perhaps, encapsulated by the report on *Tom Thumb the Great; or Harlequin King Arthur and the Knights of the Round Table*, produced on Boxing Night 1871. This is 'emphatically a Pantomime for children' claims the reviewer, 'the children so loved by the author, who crowd about him at Christmas-time, and rifle his pockets of all the good things he has hoarded up for them'.[39]

The author assumes the role of beneficent grandparent and Father Christmas, the provider and giver of treats to be discovered by the child who must claim them from him. The centrality of the child as both spectator and participator (in reality as well as in memory) locates the Drury Lane performance into a frame where the child becomes the 'lens' through which adults experience the pantomime, even on the newspaper page.[40]

Pantomime also allowed reviewers to reflect upon the apparent 'classlessness' of the audiences who thronged the London theatres, and to excuse otherwise inappropriate behaviour, or because of the influence of the season and the magical qualities of the pantomime, to note the potential to improve upon behaviour. At the Standard Theatre in Shoreditch, the reviewer enjoys the anomaly of 'shirt-sleeves' in the private boxes, and the fact that 'the dress circle is not above drinking its beer from the bottle [whilst] the gallery converses with Bob in the pit'.[41] Although children are not the focus of either the entertainment or the review, this is still an audience which takes its pantomime very seriously. This enthusiasm is experienced by reviewers elsewhere, but with less indulgence. The Pavilion Theatre situated on the Mile-end road, says the reviewer (1875):

> is not very famous for quiet audiences, but that of Friday was one of the noisiest we have seen for some time, and the roar of approval which greeted even the turning up of the gas before the curtain rose suggested the idea of how terrible must be the expression of disapproval of this crowd if they were once thoroughly roused to anger by an unfortunate performer or manager.[42]

The imminent threat of the unruly, unwashed crowd is never far away – the classes are under scrutiny. Or rather the lower classes are. In fact, the metropolitan reviewers reserve their criticisms *not* for the performance (as in the provinces), but for the audience – rowdy, noisy or just plain stupid. As the reviewer of the Hoxton Varieties' *Ali Baba, who Opened Sesame and Shut Up Hassarac* noted – 'we have a right to complain of the Hoxtonians that they prove so thoroughly unappreciative of the truly Byronic puns, with which the piece is thickly studded'.[43]

Who read these reviews of the metropolitan shows is difficult to determine. The reviewer of the Pavilion pantomime, in declining to describe the plot of *Harlequin Puss in Boots; or, Mother Hubbard and the Comical Dog Toby*, claimed that 'we [the reviewers] tell the public which are the best, and the public must go and see for themselves, when they can judge of the author's merits as well as those of the actors'.[44] There is a sense here of the *Era* correspondent performing a public service, and his review is quite distinct in tone and content from that of, for example, the *Times*. A comparison of the *Era's* Drury Lane reviews with the equally lengthy but less effusive reports found in the *Times* marks out the *Era* as fulfilling a different function and catering to a different readership. Whilst the *Times* is appreciative of the 'book' and the performers and notes the enthusiasm of the crowded house, it lacks the professional eye and agenda of the *Era*. 'A Drury-lane pantomime is distinct from pantomime in general', the *Times* writes of *The Children in the Wood; or, Harlequin Queen Mab and the World of Dreams*. But, unlike the *Era*, it cannot say why.[45]

Notes

1. The 1870s are selected for the purpose of this exercise as decade of relative calm in the theatrical world. The text-searchable on-line run of the *Era* from 1838 to 1900 is now available on the *Nineteenth Century British Library Newspapers* http://find.galegroup.com/bncn/ (all sources accessed 30 July 2008).
2. *Era*, 29 December 1878.
3. Edgar White, 'The Romance of *The Era*', *Era Almanack*, 1918, p. 66.
4. Clement Scott, *The Drama of Yesterday and Today*, 2 vols. (London: Macmillan and Co, 1899), vol. 1, pp. 414 ff.
5. Dickens, 'Holding Up the Mirror', *All the Year Round* (29 September, 1860), p. 595.
6. Ibid.
7. Scott, *The Drama of Yesterday and Today*, p. 414.
8. Dickens, 'Holding up the Mirror', p. 596.
9. *The Sphinx*, December 1868, p. 171.
10. Dickens, 'Holding up the Mirror', p. 596.
11. *Era Almanack*, 1918, p. 67.
12. In 1874, the *Era*, published each Sunday, cost fivepence. Another theatrical weekly, the *Theatrical Journal*, smaller, less comprehensive and lacking the *Era*'s range of advertising, cost one penny in 1871.
13. 'The Theatrical World', *Chambers's journal of popular literature, science and arts*, 238 (July, 1858), p. 55.
14. *Era*, 29 December 1879.
15. Interestingly, local businesses paid sometimes large sums to be represented in pantomime scenes: 'tradesmen ... willingly pay £50 or £100 for a scene which advertises their goods – that is, a scene having a view of their premises, wherein clown and pantaloon carry out a deal of practical fun with the articles in which they deal'. *Chambers's journal of popular literature, science and arts*, 260 (December, 1858), pp. 413–16.
16. *Era*, 3 January 1875.
17. *Era*, 28 December 1879.

18. *Era*, 4 January 1874.
19. Ibid.
20. *Era*, 14 January 1877; 12 January 1873.
21. *Era*, 2 February 1873. On 22nd January 1873, the *Northfleet*, an emigrant ship bound for Australia and taking railway workers and a cargo of iron to build the Tasmanian railway, was anchored in the English Channel when it was struck by a Spanish steamship. 320 people died, including Captain Knowles. References to the story of 'the Claimant' and parodies of his portly figure featured in many provincial and metropolitan pantomimes. It was a topical reference in the 1870s, since on 2 February 1874, after a trial of over 6 months, Arthur Orton was found guilty of perjury and sentenced to fourteen years' imprisonment. See Rohan McWilliam, *The Tichborne Claimant. A Victorian Sensation* (London: Hambledon Continuum, 2007).
22. *Era*, 27 December 1874.
23. *Era*, 3 January 1875.
24. *Era*, 2 January 1876; 30 January 1876; 25 January 1874.
25. *Era*, 3 January 1875.
26. *Era*, 18 January 1874; 4 January 1874.
27. *Era*, 3 January 1875; 4 January 1874.
28. *Era*, 11 January 1874.
29. *Era*, 19 January 1879.
30. *Era*, 29 December 1878.
31. *Era*, 31 December 1871.
32. *Era*, 9 February 1879.
33. *Era*, 31 December 1876.
34. *Era*, 20 December 1874.
35. 'A Children's Pantomime', *Era*, 7 January 1877 Compare with the review, quoted earlier, of *Little Jack Horner* at the Queen's Theatre, Manchester which the correspondent notes has an audience 'patronised so largely by ladies in the best circles of Manchester society' (25 January 1874).
36. *Era*, 31 December 1871.
37. See 'Mysteries of the Pantomime', *London Society*, 1:2 (March, 1862), pp. 168–74 and 'About the Pantomime', *Chambers's journal of popular literature, science and arts*, 260 (December, 1858), pp. 413–16.
38. 'A Children's Pantomime', *Era*, 7 January 1877.
39. *Era*, 31 December 1871.
40. See Anne Varty, *Children and Theatre in Victorian Britain: 'All Work, No Play'*, (Basingstoke: Palgrave Macmillan, 2008), pp. 138–47.
41. *Era*, 3 January 1875.
42. *Era*, 28 December 1873.
43. *Era*, 4 January 1874.
44. *Era*, 28 December 1873.
45. *The Times*, 27 December 1872.

Part 4

The Legacy of Victorian Pantomime

11
Continuity and Transformation in Twentieth-century Pantomime

Millie Taylor

The modern form of pantomime evolved in the late 1890s and early 1900s and owes a great deal to the Drury Lane pantomimes of J. Hickory Wood and Arthur Collins. George M. Slater wrote an author's note in his 1916 pantomime *Boy Blue* that suggests that 'nowadays the ideal of the pantomime author and producer is only to be attained by the coherent exploitation of a story in which magic and fairy powers are incorporated with humorous interludes (if possible germane to the plot) whilst endeavouring "to point a moral and adorn a tale"'.[1] This is remarkably similar to the sentiments expressed by contemporary writers and producers. Chris Lillicrap begins writing with 'a strong story', but he argues that the ideal mix is that there is something, whether verbal comedy, slapstick, variety, song or dance, for all the family.[2] Producing the right combination is a difficult juggling act for, as Gerald Frow remarks:

> The balance is a fine one and it shifts a good deal from generation to generation, production to production. All too frequently it lurches in the direction of the star or stars, but on occasion it lurches equally fatally in the direction of the story, to provide a singularly dull evening devoted entirely to what is pretty, whimsical and fantastical.[3]

In plot and character pantomime has remained very close to its late-Victorian roots. Norman Robbins (1992) and V. C. Clinton-Baddeley (1963) have traced the development of the pantomime versions of several stories identifying constant features that have come to be regarded as part of the pantomime 'tradition'. The contemporary version of *Aladdin* can be traced back to a combination of *Aladdin and his Wonderful Lamp: or the Genie of the Ring* by H. J. Byron (1856) and Byron's burlesque version of *Aladdin* (1861).[4] The pantomime version of Cinderella can be traced to Byron's 1860 'Fairy Burlesque Extravaganza' *Cinderella or the Lover, the Lackey and the Little Glass Slipper*.[5]

There are also similarities in character types across the whole range of pantomimes so that, for example, the Dame's character, whether she is a nurse, governess, cook or mother, is essentially the same. She is hard-working, down-to-earth, poor and loving towards her children, but also sex-starved, outrageous and with an adventurous spirit. The same consistency is true of other characters so that throughout the twentieth century one might expect the story to be played out between Dame, comic, principal boy, principal girl, fairy and demon who have clear characteristics and functions. The struggle between the forces of good and evil, a crucial feature of Victorian pantomime, still survives. Other supporting characters include a comic double act often loosely referred to as The Broker's Men,[6] which is their title in *Cinderella*, who have the same function in all pantomimes; they act as a catalyst to plot events and provide the opportunity for knockabout comedy. Alongside these principal characters driving the plot are singers and speciality performers. Some of the casting has changed in some productions; sometimes the principal boy is played by a man and very occasionally the Dame by a woman, sometimes there is a stepmother in *Cinderella* that alters the relationships between sisters, mother, father and Cinderella, and in particular productions the performers might be magicians, singers or comics, but the key characters and story events are remarkably consistent.

George M. Slater's *Sinbad the Sailor: An old tale retold* was performed in Birmingham in 1914, clearly giving the impression of a continuous tradition in its title. By 1992 it was Roy Hudd who was providing this sense of continuity with *Aladdin and his Wonderful Lamp: An old tale retold* performed at Watford Palace Theatre. A similar sense of nostalgia is evoked in the title *Aladdin: A traditional family pantomime* written by Chris Jordan (2002). As Jacky Bratton's essay in this volume demonstrates, part of the nostalgia of pantomime has always been the belief in a golden age of the past when pantomime had its heyday. Derek Salberg documents this through a series of newspaper reviews and articles from 1822 to 1981 that suggested that pantomime at whatever time of writing was less entertaining or popular than it used to be.[7] Gyles Brandreth continued this theme in an article for the *Sunday Telegraph* in December 2000.[8] This sense that pantomime used to be better and that it is a ragbag combination of whatever is popular contributes to the sense of nostalgia, but is also a feature both of its continuity and its transformation.

The continuous performance of pantomime throughout the country during the twentieth century is set against the background of changes in the theatre estate and the funding of theatre. While the commercial sector has continued throughout the century, first repertory theatres and then subsidised theatres developed alongside it. The idea of repertory theatres developed at the end of the nineteenth century, and the earliest companies espousing the idea of a permanent ensemble 'maintaining a broad programme of work rotating throughout the season' established themselves in Manchester, Bristol and Glasgow early in the twentieth century.[9] Progress was impeded but continued through the First World War, the depression and the coming of cinema. However, during the Second World War the Council for the Encouragement of Music and the Arts (CEMA), which had been established in 1940, began to support tours and regional ventures and was then remoulded into the Arts Council in 1945.[10] In 1965 the 'Housing the Arts' scheme was introduced, which led to a large building programme during which regional theatres were built around the country and ran predominantly as repertory theatres. By 1969 Baz Kershaw notes that there were at least 52 of these subsidised repertory companies.[11] These new theatres had auditoria of approximately 500–800 seats, which, though appropriate to the size of audience they might attract in towns across the country, is not an economically viable size when considering costs of pantomime production against income generation.

By 1989, despite the reform of the economy and the increasing requirement that subsidised theatres find business-sponsorship, grants to the subsidised theatres totalled around £40 million.[12] However, the regional theatre estate was in decline and many theatres began to reduce repertory seasons. A small upturn in the state of subsidised theatres followed as a result of urban regeneration projects and lottery funding, and the business model of smaller theatres changed to include more touring shows and West End transfers. Nonetheless, a number of regional theatres have closed, and some of those that remain are in a poor state.[13] At the time of writing pantomime in the commercial sector is dominated by a small number of large commercial producers who mount many shows and tour them to the largest commercial theatres in different towns and cities from year to year. Smaller producers mount shows in the smaller theatres, with smaller overall audience sizes, but again use economies of scale as they tour a production to perhaps 4–6 venues. Consequently, they have smaller amounts to spend on production costs. The subsidised theatres also mount pantomimes and Christmas shows which still draw their largest audiences of the year, but which have much smaller budgets than those in the commercial sector.[14]

The economics of pantomime are based on three factors: the costs of production; how the costs can be recouped; and where the profit goes. These factors continue to affect the decisions of theatre managers as to whether they choose to host a touring company or create an indigenous pantomime. The principles that govern the economics of pantomime have not changed in the course of the last century despite considerable diversity in theatre size and the introduction of a subsidised sector. Tracy C. Davis identifies the issue that there is a cyclical demand for pantomime followed by an immediate depreciation as the pantomime finishes its season in a particular location.[15] This means that a production has a finite run and consequently a finite income at each location, based on seat numbers and audience size, which gives rise to several patterns of behaviour.

First, the pantomime can tour from one year to the next but needs to be stored in the interim. Subsequent productions may be adapted, and it is rare that all the same performers will be involved as the production tours. These adaptations and the need for topicality are likely to lead to the creation of new scenes, songs or acts, usually provided by the same writers, directors and producers. This has the economic benefit of allowing producers to recoup costs over a number of years. It also leads to similar practices and techniques in the performances being disseminated across the country.

Second, the show can be broken up and separate parts of the production sold or hired out and placed into new productions. The fact that pantomime is essentially a composite of diverse parts united by a well known story, means that the stock can be sold piecemeal rather than as a whole. Smaller production companies and some subsidised companies use a combination of sale, hire and storage. However, there are examples of other types of continuity, as performers, writers and directors become attached to particular theatres. Examples of this practice include Berwick Kaler who has written the pantomime and played Dame at York Theatre Royal for over thirty years, and who has established a local following and a core team of performers who return each year.

Third and this is the most costly strategy, the set can be repainted and a new show with predominantly new performers created each year. Here, the theatre will tend to develop a 'house style' that is also likely to depend on the continuity of a particular writer or director, as at the subsidised theatres, such as Salisbury Playhouse and The Nuffield Theatre,

Southampton. The costs of production are necessarily cut to the bone at these smaller theatres.

In the commercial sector where shows are re-produced over a number of years at different venues, the writer is not always credited; the producer may be the writer or will employ writers to maintain a particular style. Here the production company and its staff of regular writers, directors, producers and choreographers ensure consistent practice that is disseminated across the country. This practice was already in place at the start of the century. *Puss in Boots* appeared at the New Theatre Cardiff in 1911 before arriving at the Prince of Wales Theatre in London in 1913 (anon). Howard and Wyndham produced pantomimes at between one and three theatres in Scotland every year from 1888 to 1948. Their empire spread to northern England where they were responsible for pantomimes at Nottingham (1924–1927), Newcastle (1930–1948), the Opera House Manchester (1934–1940), and the Royal Court Liverpool (1938–1948).[16] The directors / producers of all these pantomimes were remarkably consistent, being dominated by F. W. Wyndham, A. Stewart Cruikshank, Julian Wylie and Emile Littler with a small number of other contributors. Several of these directors also wrote the pantomimes they directed, and wrote, produced or directed at other theatres too.

Derek Salberg notes that from the mid-twenties onwards those towns that had produced their own pantomimes were gradually less inclined to do so, and brought in productions from the large scale producers.[17] Since then many production companies have continued to grow and have developed through a process of takeover and amalgamation.[18] E&B (developed from Triumph) was the largest company in the 1990s and moved productions between Plymouth Theatre Royal, Birmingham Hippodrome and up to 30 other venues. They amalgamated with Nick Thomas and John Conway to become QDos and produced 21 pantomimes in 2008. Paul Holman Associates and First Family Entertainment each had 12 productions in 2008. Hiss & Boo and Evolution are smaller commercial producers who mount 4–6 shows in smaller theatres and some subsidised or regional theatres. For example, Evolution mounts shows at the Yvonne Arnaud, Guildford, which has a small subsidy from the Arts Council, discontinued in 2009.[19] This has been a subsidised regional theatre buying in a small commercial pantomime, though it has always had a particularly commercial outlook and links to the commercial sector mounting many successful tours and west end transfers. These smaller producers tend to use a relatively small number of writers and directors, who may also be performers, and so the traditions continue.

These production companies have different contractual arrangements with theatre managers about the way the costs and profits are allocated, but the basic principles have not changed. In this commercial sector it is likely that theatre managers would take a share in the profits. The producer would have the outlay and the largest share of the risk, but would also take the largest profit from the production. In interviews conducted in 2003–4 the largest producer reported that the company could afford to spend up to £400,000 on a new production and recoup the costs over ten years. Each year they might spend £20,000–£30,000 on refurbishment and £10,000–£15,000 on new scenes so that the show was always fresh and topical for the venue, and was adapted to suit the abilities of the star performers it employs. The consequences are that similar practices, routines, music and designs appear around the country leading to a remarkable continuity of practice and a slow pace of change through the century.

The second practice, where new shows are required annually at relatively small venues or groups of venues, and which is characterised by writers, performers and directors having an association with a particular venue, was apparent at the start of the century. J. Hickory Wood and Arthur Collins created the spectacular pantomimes for Drury Lane from 1900 onwards.[20] George M. Slater was another key writer of the first decades of the century. His pantomimes for Birmingham Theatre Royal include *Sinbad the Sailor* (1914), *The House that Jack Built* (1915) and *Boy Blue* (1916) and for Drury Lane *Dick Whittington* (1932). Examples of this type of association in the recent past include Berwick Kaler at York, as mentioned above. Roy Hudd has written and performed for E&B Productions and later for Hiss & Boo, which had a successful connection with Queen's Theatre Barnstaple from the mid 1990s. Chris Denys and Chris Harris had an association with Bristol Old Vic for several years and Chris Jordan writes a new show each year at Eastbourne Theatre. Here the name of the writer contributes to the continuity of the production.

Alongside the association between theatre and writer and/or performer is the requirement to produce a new show every year. This need to innovate has led to slightly more diversity, for example, in the stories of pantomime and in the comic routines.[21] Tracy Davis records that from the mid-Victorian period London entrepreneurs sold off pantomime stock, and that the re-use of 'pantomime accoutrements' was routine by the early twentieth century.[22] The practice has continued through the century, especially in relation to particular scenic effects or sets for comedy routines. A wonderful flying horse and coach for the Act One finale of *Cinderella* has appeared at the Northcott Theatre Exeter (a subsidised theatre), and at Eastbourne Theatre (a small civic theatre that runs as a touring house during the year but produces its own pantomime). A spectacular set for a comedy bathroom scene has been seen at York Theatre Royal in *Mother Goose* in 2003–4 and has subsequently appeared at a number of other places. It is likely to provide the basis for the reworking of a set-piece comic routine at the Yvonne Arnaud Theatre, Guildford in the 2010–11 pantomime. At this scale the pantomime is generally the subject of an ongoing negotiation between theatre manager and producer who are more likely to share the risks of mounting the show and the share of profits.

The third practice, of repainting and re-using sets from one year to the next, and cutting costs to the bone, is particularly apparent in subsidised theatres. Although the costs are much smaller these theatres still struggle to make a profit on their annual pantomimes, which can be used to underpin other works in the season. Pantomime does not always turn a profit, though, as Tracy Davis records. St Helen's Theatre Ltd made a loss on the pantomimes in 1904–5 and 1905–6, but despite this, 'the theatre was kept afloat by a very narrow margin and ... the pantomimes gave a measure of security that nothing else afforded'.[23] Salisbury Playhouse spent £73,000 on their 2003 production of *Aladdin* and recorded a profit of between £40,000 and £50,000 which helped to fund the rest of the season. Kerry Michaels at Stratford East reported that because of the low ticket prices and the many concessions the theatre offers, pantomime makes a loss. However, it is usually the smallest loss of the year, and the pantomime is the most successful show of the year bringing in new audiences. These productions, in what are still subsidised repertory theatres, tend to be geared to their local audiences, which in Stratford is ethnically diverse, while in Salisbury it is still predominantly white and middle class. Such radically different contexts affect music, dancing, characterisation and topical allusions at these theatres, which have a more diverse approach than the pantomimes in the commercial sector.

The costs have risen to the extent that there has been a substantial reduction in the number of performers and crew employed for pantomime during the course of the century. As theatres were replaced by cinemas in the 1930s there were fewer pantomimes and smaller audiences for them. The Drury Lane pantomimes in the first decades of the twentieth century saw the most spectacular scenic effects and the employment of hundreds of extras. A production of *Puss in Boots* that transferred to the King's Theatre Edinburgh in 1918 had a cast of 150. The following year 170 artistes appeared in the 1919 production of *Babes in the Wood* at the same venue.[24]

Immediately after the Second World War the numbers employed dropped substantially, but Norman Robbins still records a chorus of twenty-four-plus juveniles and a cast of sixteen with a twelve-piece orchestra.[25] Since then there has been a further steady decline. In the commercial sector, E&B's production of *Mother Goose* in Aberdeen in 1989 had a cast of ten led by Matthew Kelly. These were supported by six dancers and a group of juveniles with eight players in the band. At the Yvonne Arnaud Theatre, Guildford (a much smaller theatre but still an E&B production), for *Aladdin* in 1992, the band was reduced to five players, with eight principals, five dancers and children. At this time the company was reducing dancers and musicians who were on minimum union contracts, in order to retain the services of more highly paid stars, who were expected to attract audiences. Jim Davidson's commercial productions sometimes dispense with live music altogether.

During the 1990s the small subsidised theatres have reduced the numbers employed even more drastically because of the economic limitations identified above. A small cast is often supported by a chorus of children. Some have maintained a band of three, but gradually with the increased use of synthesisers and then computers/sequencers this is sometimes reduced even further. Occasionally, there might be only one musician employed, and some theatres use actor-musicians in a new style of pantomime performance.[26] Scenery is reused and some special effects or costumes are hired for the performance. Backstage staff numbers are severely limited, reducing the capacity to clean up after slosh scenes and undertake transformations. The result of the reduction in the number of dancers and the decline of scenic extravagance is that smaller pantomimes are transforming into a new breed of interactive musical comedy, without the choreographic artistry or physical comedy seen in costlier productions. Some of the very largest commercial productions are heading in a similar direction for different reasons with extended musical and vocal numbers performed by experienced singers in the principal boy and girl roles.

Productions in the commercial sector are still built around the abilities of the stars and geared to their skills, in line with the practice that existed earlier in the late-Victorian period and has been apparent throughout the century. Music hall, musical comedy and variety performers have at different times transformed the content of the pantomime by introducing their skills and their acts. The Victorian introduction of music-hall performers was followed by the incorporation of musical-comedy stars in the early twentieth century and then variety performers made famous by radio and television, some of whom continued to appear live in pantomime, especially in comic and Dame roles, into the twenty-first century. These variety performers included Charlie Cairoli, Les Dawson, Morecambe and Wise, Dick Emery, Jack Tripp, Roy Hudd, Arthur Askey, Billy Dainty, John Inman and Jack Douglas. Through the impact of television these variety performers have arguably had the greatest influence on the development of pantomime comedy in the course of the twentieth century. This was not only because of their ability to attract to the theatre the family audiences tuning in to light entertainment on radio and television, but also because of the possibility that their comic routines and physicality might be recorded on film and television, and so influence other performers and creators.[27] However, alongside the variety and comedy stars, singers were also being introduced in the 1950s and 1960s from pop music. Peter Holland and Gerald Frow between them list Frankie Vaughan, Cliff Richard, Frank Ifield and Engelbert Humperdinck amongst those who played principal boy.[28] But this is not a simple linear process of transformation, since music hall singers had already appeared in nineteenth-century pantomime and sports stars have appeared in pantomime in both centuries. Today each pantomime contains performers of different ages and from different areas of celebrity (music, television, sport, and more recently, soap operas and reality shows) employed to attract the widest possible audiences.[29]

The economics of producing slapstick and slosh scenes with performers who no longer bring their own discrete acts requires more rehearsal time. It also requires more costumes and more backstage staff, and is costly to produce. As a result of the enormous growth of employment opportunities in musical theatre since the late 1970s performer training has focused on the ability of young performers to perform in musical theatre rather than in comedy. So as the apprenticeships through variety and music hall into pantomime and summer season that were apparent for over half of the century have gradually disappeared, and as the older performers retire thus reducing the apprenticeships within pantomime itself, pantomime has begun to alter to incorporate the skills of the new performers. It is now becoming focused towards musical theatre, singing and dancing, and stand-up comedy. The consequence is that apart from the designated comedians many commercial pantomimes are moving towards musical comedy with quite long opening sections of song, dance and underscoring of warm-up comedy spots.

The way the musical score is structured has not changed much; throughout the twentieth century the music for commercial productions has been largely compiled from well-known sources. In twentieth-century scripts there are references to the songs that might be used for particular characters and to composers or musical directors who would compose incidental music such as 'tinkly and mysterious music'[30] for entrances, exits, scene changes and so on. However, the sources and styles of this music have changed. In 1907 J. M. Glover is credited with composing, selecting and arranging the music for *Babes in the Wood* at Drury Lane, which contained a combination of popular and classical repertoire.[31] By 1990 the Guildford production of *Robinson Crusoe* consisted of a combination of music hall and musical comedy songs, pop music, music from Disney and other films, themes from soap operas or other television programmes, and incidental music. The music was played by a band of five consisting of electric keyboards, electric guitar and electric bass, triple wind player (fl, cl, sax) and trumpet. The opportunities for spectacular scenic design and extended dances and tableaux have been reduced in the course of the century because of the reduction in performers and the costs involved, but the instrumentation of the band also means that extended musical intermezzi are less effective. Overall, while pantomime music has been compiled and structured in the same way throughout the century, the content of the music has been transformed to reflect current tastes and the size and instrumentation of the band.

The songsheet is now one of the most important musical items in a pantomime because of the opportunity for audiences to join in, but it is not clear when the current practice began. At the start of the twentieth century audiences were accustomed to being demonstrative and joining in with the choruses of songs or responding to comedians in Music Hall and later Variety.[32] The practice of audiences expressing their views, shouting and booing continued through into the 1950s, when gradually audiences at many theatre performances were encouraged to become more restrained.[33] In pantomime, however, audiences continued to join in with well-known songs, and opportunities to interact with the performance have remained. The received wisdom is that, as finale sets are the most spectacular and elaborate, a longer time is required to construct them and so the songsheet appears before the finale to provide this time. I have not found any specific reference to the songsheet at this point in the show in the early part of the century. In J. Hickory Wood and Arthur Collins' *Babes in the Wood* (1907) the script suggests that the final scene followed directly from the previous scene and contained a couplet for each of the principal characters. By 1952, there is a song sheet, 'Gin put the ginger in Grannie', before the final scene in

Cinderella.[34] There may be a correspondence between the reduction in stage crews and the introduction of the songsheet and its competitive element to cover the time and the noise of scenery shifting. It is possible, therefore, to speculate that the specific identity and function of the songsheet has evolved in the course of the century out of the old practices of audiences being encouraged to join in with choruses in music hall, variety and Victorian pantomime.

There are other moments of interaction in contemporary pantomime, such as 'It's behind you' and 'Oh no it isn't'. Norman Robbins records that 'Oh no it isn't' began in a variety act known as 'The Two Pirates'.[35] The two performers pretended there was no wire for their balancing act. The larger performer remarked to the audience 'There's no wire you know'. He usually got a response, at which point he would address the audience with 'Oh no there isn't' to which the reply was the now ubiquitous, 'Oh yes there is'. Other opportunities for interaction are stipulated in the Dame's and comic's opening spots which are written out in some early scripts, though doubtless many were extended or improvised. The point is that, although pantomime was already interactive at the start of the twentieth century, some of the particular events of interaction that are now 'traditional' are more recent and stem from particular situations or performers.

While the numbers employed in pantomime have declined, the ability to produce effects through the use of technology has developed. Julian Wylie worked with Strand Electric in the early 1930s on the development of Ultra Violet lighting.[36] Derek Salberg refers to the use of Ultra Violet for a scene in his 1956 pantomime *Babes in the Wood* remarking that it had been used many times both before and since. It is still extremely popular, especially for underwater scenes in which wonderful effects can be produced using very limited human resources. Examples of this are the underwater scene in *Sinbad the Sailor* at York Theatre Royal in 2007 and a UV ballet at the top of the beanstalk in *Jack and the Beanstalk* at London's Barbican Theatre in the same year. [37]

Film is also used to create easily reproducible scenic effects, especially chase sequences. The presence of a screen downstage also allows upstage scene changes, and the filmed material can replace physical chases around the auditorium. Norman Robbins documents a very early example of this in a 1913 production of *Robinson Crusoe* in Exeter. In this a film of 'all the denizens of the sea could be projected whilst an aeroplane would be seen flying overhead'.[38] Derek Salberg used a car chase film in 1961 in a production of *Aladdin*.[39] More recently, in 1990, a similar car chase was used in *Robinson Crusoe* at Guildford's Yvonne

Arnaud theatre. 3D film sequences are now a feature of the largest commercial productions.[40] This is a strategy predominantly used by the commercial producers who can recoup the outlay over a number of years. At York Theatre Royal a short film is often inserted into the pantomime but this has to be shot each year to contain the new characters and costumes, and to reflect the theme of the show. York is, therefore, the exception among smaller companies.

Lasers and computerised lighting effects have been introduced for special effects, but the most notable technological development in what is generally a fairly 'low-tech' production is in the use of microphones and electronic instruments. Pantomime has always been a noisy affair. At the start of this volume Jim Davis remarked on the rowdy pantomime audiences that sometimes completely drowned out the performers. By the 1950s pop singers required microphones to reproduce the sound of their rock'n'roll records, but as the instrumentation of orchestras was exchanged for small bands of brass and electronic instruments microphones were needed for all singers. Click tracks became common in the large commercial productions to supplement the live singing, especially of dancers, and sometimes to augment the sounds the band could make. By the 1990s all speaking performers, who were no longer accustomed to projecting their voices or controlling audiences in large theatres, expected to use a radio mike for dialogue as well as songs. This means that audiences can be stirred up to interact vocally with the stage, and performers can still (mostly) be heard and maintain control of this vocally interactive audience.

Two things appear to have occurred simultaneously: the economics of production has led to a reduction in musicians and so the volume and energy of the music is produced by electronic amplification of a small band rather than by the larger orchestra; secondly, the use of microphones and electronic instruments and the transformation in popular music have inspired a change in the musical content of pantomime. This must mean that pantomime now sounds different, but importantly, the balance between performer and stage in terms of volume, interaction and control has changed. In the larger theatres the performer has foldback speakers to hear the orchestra and her voice, and so is less likely to hear audience commentary. Meanwhile at some performances the sound level is amplified so much that audiences can talk and shout without being heard over performers, so any interaction with the stage is likely to be no more than formulaic. Baz Kershaw describes the change in power relations between audiences and performers in the 1960s in other types of performance.[41] In pantomime, the interactions have remained but technology has rendered them all but meaningless in the larger theatres.

Local and topical references are also a continuous feature of British pantomime, and are an obvious site for gradual transformation as they refer to current events of contemporary interest as well as identifying local features. This practice was common in the nineteenth century as the essays by Davis, Foulkes, Sullivan and Norwood have indicated. They give the large impersonal productions the opportunity to appear to relate to local communities. In *Puss in Boots* of 1913 Wiggles, the comic, attempts to persuade the Count to invest in wireless telegraphy. In the 1929 Drury Lane production of *The Sleeping Beauty* the Good Fairy and the Prince arrive in Meccanoland where the children build Prince Florizel an aeroplane to go and rescue the princess. At the end of the century mobile phones and aeroplanes make appearances and there are frequent references to local landmarks, the difficulties encountered on public transport, in traffic jams and in parking.

Contemporary events and especially politics have been the source for comic one-liners throughout the twentieth century. In *Puss in Boots* (1913) there is a sideswipe at parliament with the comment: 'What is parliament Why one house a night and a rotten programme'. In 1973 Frow records that Widow Twankey bemoaned the fact that she was getting fewer brassieres at the laundry every week, blaming women's liberation,[42] and in 1983 the Dame sang about being on the Dole.[43] Pantomime also retains the freedom for ad-libs to be inserted. In recent years these have included references to the disappearance and discovery of Saddam Hussein, Peter Mandelson's resignation and Cherie Blair purchasing a flat in Bristol all inserted as one-off one-liners at the time they hit the news.[44]

Throughout the twentieth century pantomime has referred to other theatre forms. In *Puss in Boots* 1913 and *Sinbad* 1914 there is reference to Black-Face minstrelsy, and in both cases these are cues for speciality minstrel spots. In *Sinbad* (Slater 1914) Tinbad and the Caliph discuss going to the Halls. By the twenty-first century equivalent references were predominantly to television, film and pop music. In Joanna Reid and Stuart Thomas's *Aladdin* (2003) there is a series of references to *XMen*, *The Matrix*, *A Fistful of Dollars* and *Dixon of Dock Green*. The importance of these references is that, especially for the smaller and poorer productions, witty topical references and other forms of reflexivity can replace physical comedy in creating enjoyment for audiences at a fraction of the cost.

Gerald Frow remarks that he believes the greatest change in pantomime of the twentieth century is the amount of audience participation.

'It is, after all, the only form of entertainment which still permits its audience the pleasure of joining in, and in many pantomimes the company keep up an almost constant running dialogue with the "customers" more redolent of a music-hall than any legitimate theatre.'[45] Whether this is really a transformation from performances at the start of the century is debatable and possibly difficult to determine, and the power relations of this exchange have almost certainly been transformed by technology. What can be deduced is that there is a continuity of structures, characters and practices alongside (and despite of) the transformation of some of the content so that pantomime may remain contemporary in its comic references, its music, in the use of stage technology and in its responsiveness to a changing economic climate.

Scripts consulted

Anon. *Puss in Boots* New Theatre Cardiff, 1911, and Prince of Wales Theatre, 1913.

Anon. *Aladdin* Yvonne Arnaud Theatre, Guildford: E&B Productions, 1992.

Chissick, Jack and David Horlock *Jack and the Beanstalk* Salisbury Playhouse, 1987.

Clinton-Baddeley, V. C. *Cinderella: or Love makes the world go round* (London: Samuel French, 1952).

Davies, Tudor. *Robinson Crusoe* Yvonne Arnaud Theatre, Guildford: E&B Productions, 1990.

Denys, Chris and Chris Harris *Dick Whittington* Bristol Old Vic, 1995 (published London: Josef Weinberger, 2001).

Denys, Chris and Chris Harris *Jack and the Beanstalk* Bristol Old Vic, 1994 (published London: Josef Weinberger, 2002).

Elliott, Paul and Tudor Davies *Dick Whittington* Yvonne Arnaud Theatre, Guildford: E&B Productions, 1991.

Hickory Wood, J. and Arthur Collins *Babes in the Wood* Drury Lane, 1907.

Hudd, Roy *Aladdin and his Wonderful Lamp: An old tale retold* Watford Palace, 1992.

Jordan, Chris *Aladdin* Potter's Bar, 2002.

Jordan, Chris *Cinderella* Devonshire Park, Eastbourne, 2003.

Maurice, Newman *Dick Whittington* Lyceum Theatre, 1919 (produced by Walter and Frederick Melville).

Morley, John *Dick Whittington* (London: Samuel French, 1986).

Reid, Joanna and Stuart Thomas *Aladdin* Salisbury Playhouse, 2003.

Robbins, Norman *Hickory Dickory Dock* (London: Samuel French, 1979).

Slater, George M. *Sinbad the Sailor: An old tale retold* Birmingham Theatre Royal, 1914.

Slater, George M. *The House that Jack Built* Birmingham Theatre Royal, 1915.

Slater, George M. *Boy Blue* Birmingham Theatre Royal, 1916.

Slater, George M. *Dick Whittington* Theatre Royal Drury Lane, 1932.

Notes

1. George M. Slater, *Boy Blue* Birmingham Theatre Royal, 1916.
2. From an interview in 2003.
3. Gerald Frow, *'Oh, yes it is!' A History of Pantomime* (London: British Broadcasting Corporation, 1985), pp. 181–2.
4. V. C. Clinton-Baddeley, *Some Pantomime Pedigrees* (London: The Society for Theatre Research, 1963), p. 31
5. Ibid., pp. 10-11.
6. They are the Chinese Policemen in *Aladdin*, the Robbers in *Babes in the Wood*, they may be Captain and Mate or aides to King Rat in *Dick Whittington*
7. Derek Salberg, *Once upon a Pantomime* (Luton: Cortney Publications, 1981).
8. Gyles Brandreth 'Oh yes it is!' *Sunday Telegraph Review* December 10, 2000, pp. 1–2.
9. Olivia Turnbull *Bringing Down the House* (Bristol and Chicago: Intellect, 2008), p. 18.
10. Ibid., pp. 22–7.
11. Baz Kershaw (ed.) *The Cambridge History of British Theatre Volume 3, Since 1895* (Cambridge: Cambridge University Press, 2004), p. 301.
12. Ibid., p. 311.
13. For more detail see Kershaw *The Cambridge History*, pp. 316–25 and Turnbull *Bringing Down the House*.
14. These patterns of economic behaviour are not discrete with overlaps between commercial and subsidised theatre practices depending on individual circumstances and policy.
15. Tracy C. Davis *The Economics of the British Stage 1800–1914* (Cambridge: Cambridge University Press, 2000), p. 349.
16. www.its-behind-you.com/howardwyndham.html accessed 14.05.07.
17. Salberg, *Once upon a Pantomime*, p. 12.
18. The economics of production of pantomime in repertory and commercial theatre over the past twenty years is discussed in more detail in Millie Taylor *British Pantomime Performance* (Bristol: Intellect, 2007), pp. 21–32.
19. Turnbull *Bringing Down the House* pp. 164-5.
20. Frow, *'Oh, yes it is!'*, p. 168.
21. At York Theatre Royal, which is a subsidised theatre, *Dick Turpin* in 2008 was a sell-out pantomime and was recorded for the Victoria and Albert Museum Theatre Collection archives.
22. Davis *The Economics of the British Stage*, p. 346.
23. Ibid., p. 343.
24. (www.its-behind-you.com/howardwyndham.html accessed 14.05.07). Tracy Davis records the employment of 5640 backstage labourers in London in 1911, and 2054 in the same year in Manchester, though this is for an April census and not at Christmas when it is possible even more staff were employed pp. 314–15.
25. Robbins, *Slapstick and Sausages: The Evolution of British Pantomime* (Tiverton: Trapdoor Publications, 2002), p. 200.
26. This has been seen at Theatr Clywd and Winchester Theatre Royal among others.

27. For example, the Mirror Routine has continued to be performed in pantomime, most notably recorded by Jack Tripp and Roy Hudd and archived at the V&A Theatre Collection. However, it also provides the comic scene in a Marx Brothers' film and in an episode of French and Saunders. As Mayer suggests much more research needs to be done on the interactions between pantomime and recorded forms.
28. Peter Holland, 'The Play of Eros: The Paradoxes of Gender in English Pantomime' *New Theatre Quarterly* 13:1 (1997), p. 197 and Frow, *'Oh, yes it is!'*, p. 183.
29. In a 1990 pantomime, *Robinson Crusoe* at the Yvonne Arnaud Theatre the cast included: a children's television performer, an astrologer and TV presenter; a seasoned pantomime veteran, a sports star – Olympic medal winner Tessa Sanderson – a presenter from a light entertainment television show and a variety performer.
30. V. C. Clinton-Baddeley, *Cinderella: or Love makes the world go round*. (London: Samuel French, 1952).
31. Songs include 'Money Boy' which is adapted from 'Honey Boy', while Marion sings to Leoncavallo's 'Barcarolle'. The music for the spectacle The Passing Hours (Scene 6) is drawn from the Entr'acte to the *Tales of Hoffman* (Offenbach), followed by a piece by Chaminade, a piece by Saint Saens interspersed with dances composed by the musical director.
32. Baz Kershaw (ed.), *The Cambridge History of British Theatre*, p. 299.
33. Dan Rebellato notes that this represented a 'class struggle' in the literary theatre, as the power was transferred from audience to performance during the 1950s and 1960s, but that pantomime audiences remained vocal. *1956 And All That* (London and New York: Routledge, 1999), pp. 106–7.
34. Clinton-Baddeley, *Cinderella: or Love makes the world go round* (1952).
35. Robbins, *Slapstick and Sausages*, p. 203.
36. Ibid., p. 195.
37. Dan Rebellato records that while the technology changed substantially throughout the century, and in the mid-century lighting designers achieved autonomy and a degree of professionalisation, their status and working conditions have not improved substantially. Dan Rebellato *1956 And All That,* pp. 89–94.
38. Robbins, *Slapstick and Sausages*, p. 198.
39. The plot of the routine is documented in Salberg, *Once upon a Pantomime,* p. 153.
40. For example in *Robinson Crusoe and the Caribbean Pirates* at the Theatre Royal, Newcastle Upon Tyne in 2008–9.
41. Kershaw *The Cambridge History of British Theatre*, p. 299.
42. Frow, *'Oh, yes it is!'*, p. 146.
43. In Denys and Harris *Jack and the Beanstalk* (1994) there are references to 'the Council Tax, the TV tax, the Road Tax, the Value Added Tax, the Tin Tax, not to mention the Cow Tax'.
44. Taylor, *British Pantomime Performance*, p. 137.
45. Frow, *'Oh, yes it is!'*, pp. 185–6.

12
Victorian Pantomime on Twentieth-century Film

David Mayer

This essay should be read as one long footnote. In contrast to previous chapters in this volume which offer further insights into the methods and personalities and venues of Victorian pantomime, this chapter points neither to fresh understandings nor to re-interpretations. My colleagues' valuable chapters necessarily draw directly on available, traditional, and orthodox means of research: libretti (i.e., the 'books of songs and choruses' for individual pantomimes sold in theatres as combination programmes and souvenirs), reviews, props, costumes, theatre records, architectural evidence, memoirs, anecdotes, audience data, and a miscellany of iconography. In contrast, this afterpiece turns in a different direction and, as its purpose, introduces hitherto unused, perhaps unexpected, research tools. Unapologetically, I point to archival materials[1] largely unknown to theatre historians and to the certainty of further discoveries to be made by exploring these very sources.

I am of the view that in early twentieth-century archival cinema films there is a largely untapped source for pantomime research awaiting serious and close investigation by future scholars. I do so aware that the qualifying adjectives *Victorian* and *Edwardian* instantly place pantomime outside – or just inside – the modern era and, consequently, seemingly exclude a theatrical form lost to recording and preservation through the medium of film. The idea of Victorian pantomime preserved and depicted on twentieth-century film may strike us as an oxymoron, a contradiction in terms. How can the cinema, so much a manifestation of modernism and twentieth-century technology, throw much light on the Victorian and Edwardian stage? Victoria died in 1901, Edward in 1910. Motion pictures did not offer many narrative subjects until ca. 1900. Those films which have survived from the 1890s and the very early 1900s are brief, many less than complete, often lacking their beginnings and endings and at best a few minutes long – either because portions of the entire film have been lost or, as likely, because an exhibitor bought and exhibited those segments which he assumed his patrons would most enjoy.[2] By 1908, with the introduction of industry standard technologies, surviving films are longer, some with durations of ten or twelve minutes. By the 1910s, a film with a running time of thirty-five minutes is still remarkable, and by this date the Victorian and Edwardian ages are altogether extinct – and so, too, are the Victorian and Edwardian theatres. Or so we might expect. In actuality, elements of the Victorian stage carry on for several decades beyond 1900. And further, conditions conspire to bring pantomimists and pantomime sequences before the camera well into the 1920s and for there to exist today, a full century later, a body of pantomime and Victorian pantomime-related motion pictures worthy of scholarly enquiry and spectator pleasure.

I am not the discoverer of these films. Credit for their discovery and for the myriad pleasures they impart goes, principally, to two archivist-historians: Bryony Dixon, Curator of Silent Film at the British Film Institute, who, for several years, has been gathering and cataloguing films relating to the English music hall,[3] and Mariann Lewinsky, who, in Switzerland, has been recovering the early work of the dominant film company, Pathé-Frères. Modest about the value of their findings, these two professional specialists have shown occasional treasures at European silent film festivals, but there have been fewer opportunities to present these films to theatre historians. The American film scholar, Matthew Solomon, who studies the relationships between stage and cinema conjuring and macro-illusions, has also brought forward early films which inform the *mise en scéne* pantomime and repay close study. The film programmes Solomon has curated have addressed theatrical praxis insofar as such praxis relates to stage magic, but he has stopped short of considering close – perhaps obvious – links with Victorian pantomime.[4] In these circumstances, my role has been to take a further step, offering a partial inventory of these discoverers' findings, joining their film discoveries to theatrical pantomime, and recognising the import that such films hold for theatre research.

I offer no analyses of the films I cite, merely the briefest of descriptions of the most pertinent ones and indications where these films may be found. The various kinds of pantomime films I describe cover a limited time- span. The earliest film dates from 1899; the very last of these was shot in 1924. Individually and in aggregate, they reveal much about the performance of pantomime and the expectations that this entertainment generated in theatre audiences.

These surviving films fall into several categories. Some were conceived as advertising trailers for stage productions; some, *féeries* and *diableries* and films of magic and illusion, were made on the European continent as entertainments intended for variety bills on which these brief – approximately five to ten minute – movies constituted a fraction of the overall film programme. The most heterogeneous and the briefest are those films of variety performances made or commissioned by the early manufacturers of motion picture cameras, projectors, and lenses who sought to provide ready-made, accessible, and enjoyable 'software' products as an inducement to would be filmmakers and exhibitors to purchase their expensive hardware.

Thomas Edison was such a manufacturer. Although at first resistant to the notion that projected film had much of a future, his backing for filmmaking increased between his first cautious experiments in 1890 and his full-scale step into commercial films ca. 1902. In the 'nineties and earliest years of the twentieth century, Edison's agents engaged circus performers, dancers, acrobats, music hall and vaudeville artists, and pantomimists to repeat their live stage and tent show acts on film. While his factories made camera equipment, his studios, first in New Jersey, later in the Bronx, turned out more than 900 motion pictures, many of these 'actuality' or news events, but a substantial number depicting variety performances.[5]

Such a film is Edison's *Three Acrobats* shot in March, 1899,[6] a mere 26 feet of film with a running time of less than 26 seconds.[7] Brief though this fragment is, this film depicts two male performers, one dressed as Clown, another as Harlequin, pursued by a female acrobat, dressed in tights and carrying a slapstick, through a vertical trap which alternately hinges and pivots and consequently obliges different manoeuvres to effect passage. Here we see in action – and in use – one of Georges Moynet's *trappes anglaises,* depicted, but necessarily immobile in his *Trucs et Décors,*[8] the book absolutely essential to the study of pantomime and, indeed, to our understanding of all mid- and late-Victorian stage machinery. Fortunately, this is not the only motion picture footage to show English trapwork in use, but it is, by some four years, the earliest to survive in the Library of Congress' holdings of 'paper prints'.[9]

An example of a narrative film which offers a far broader repertoire of stage effects common to pantomimes, but which appears four years later, intended to be enjoyed by music hall and vaudeville audiences, is Georges Méliès's musical burlesque *diablerie, Faust aux Infers* (1903).[10] Méliès' involvement with filmmaking falls between 1896 and 1913, but his immediate and continuing contact with English culture, especially conjuring and pantomime, began in January 1884, when, employed in his family shoe business, he was sent to London. There he saw E. L. Blanchard's *Cinderella* at Drury Lane, George R. Sims' *The Golden Ring* at the Alhambra, and *Red Riding Hood* at Her Majesty's and attended performances of magic and illusion by Maskelyne and Cooke at the Egyptian Hall.[11] Returning to Paris, obsessed with conjuring, macro-illusions, and fantasy narratives, Méliès became an avid, if not immediately accomplished, amateur stage illusionist. In 1888, selling his interest in the shoe-making plant, he purchased the conjuring theatre of Jean-Eugène Robert-Houdin where he and other conjurors gave performances of the sort he had witnessed in London, devised and staged *féeries* and musical burlesques, and exhibited automata.[12] What compels our interest in this small theatre was its capacity to mount illusions comparable to those available to any Victorian pantomime machinist or 'arranger'. All that this theatre lacked was sufficient light to enable early filmmaking and space in which to mount and move scenic tricks. Thus, in 1897, Méliès constructed his own film studio in the Paris suburb of Montreuil, comparable in the deployment of machinery to that in the Théatre Robert-Houdin but able to admit daylight and, later, strong artificial light. He continued to make films until 1913.

Faust aux Infers reveals some of the machinery of pantomime illusion in operation. As Faust enters Hell, he is escorted by Mephistopheles who points out the environment to which Faust has been condemned. As groundrows move laterally and pyrotechnics erupt, the pair step onto a slow trap which bears both characters downward. Their descent continues, the devil and the sinner suspended in flying harnesses, whilst a backcloth – a vertical diorama[13] – is unrolled in an upward direction to further create the illusion of bodies in free-fall. At the lowest level, the spectator experiences the illusion of sinking still further into Hell's rocky depths, but the illusion is actually a 'sink and rise' effect: profile groundrows sinking on 'sloats' whilst a series of backcloths and overhead profile pieces are flown upward, each parting of groundrow, overhead profile, and backcloth adding to the illusion of inward and downward movement. *Faust aux Infers*'s sink-and-rise is, to my knowledge, the only instance of this common Victorian stage illusion captured on film.[14] There are dances of *petit diablerines* (Méliès's films, in common with most English pantomimes, are remarkably prodigal in their deployment and cheerful exploitation of young females in tights. Paris theatres provided a ready source of young chorines who brought their costumes and their sometimes rudimentary dancing skills to Méliès's studio) and a chorus of devils who then gather in an apotheosis as Mephistopheles, his ascent accompanied by pyrotechnics, arises through a trap.

Méliès's films offer one of the more accessible routes into French *féeries*, a theatrical genre which covers nursery and folktales of mortals encountering supernatural creatures, enchanted dwellings, and strange events, the stuff of pantomime openings. In recent years numerous Méliès films have been restored and made available on DVDs by Lobster Films in France[15]and have been the subjects of exhibitions and significant screenings. The most notable of these was organised for the 2006 annual Pordenone Silent Film Festival by Dr Solomon who, rather than concentrate on Méliès to the exclusion of contemporary filmmakers who also dealt in fantasy and illusion, offered an entire festival segment, some twenty-four films in all, on 'Magic in Film'.[16] In addition to Méliès's *Le Parapluie Fantastique* (1903) and *Le Portrait Sprite* (1903), both films with illusions drawing upon complicated stage trapwork as well as the cinema trick of stop-action photography, Solomon showcased Gaston Velle's *Les Fleurs Animées* (1906), an orientalist fantasy, Pathé's *Les Chrysanthèmes* (1907), and another oriental piece, and Pathé's *Métamorphoses du Papillon* (1904), the transformation of a butterfly into a woman, further elaborated in *Papillons Japonais* (1908), where the transformation is from butterfly into a woman fluttering tinted butterfly wings and thence into a full Löie Fuller 'butterfly dance'.

Even more germane to the subject of this essay, Solomon introduced three English films which sought to reproduce by cinematic trickery the stage effects – vertical pivoting and revolving traps, floor traps, and falling flaps – essential to numerous Victorian pantomime Harlequinade scenes in which unsuspecting travellers find themselves in haunted kitchens where food behaves in surprising ways, hotel rooms in which beds and mirrors, washstands and wardrobes slide about, disappear and reappear elsewhere, and guarantee that rest is impossible, and mysterious coaches which collapse, ensnare their occupants, and move eccentrically. In Robert W. Paul's *The Haunted Curiosity Shop* (1901), Paul's *Undressing Extraordinary; or, The Troubles of a Tired Traveller* (1901),[17] and the Warwick Trading Company's *The Cabby's Dream* (1906), we witness the comic illusions, described and diagrammed by Moynet, but here replaced by stop-action and clever splicing. Cinema techniques – the rudimentary technology of a modernist craft – thereby sought to emulate and improve upon the sophisticated illusions of the Victorian stage. But few filmmakers were confident enough of the new tricks to abandon the old methods entirely. Whilst Solomon's programme, and in particular the English films he nominated, confirm that stage illusions and sight-gags essential to pantomime had been taken up by filmmakers, incorporated into film studio scenery, and employed in romantic fairy-tale or humorous knockabout narratives for the twentieth century's new film-going audiences, these and other contemporary films still enable the theatre scholar capable of recognising theatrical scenic practices to view standard Victorian stage devices operating alongside emerging film techniques.

Mariann Lewinsky's discoveries have recently emerged at Bologna's annual festival of recovered and restored film, Il Cinema Ritrovato.[18] In segments dedicated to exhibiting European films released exactly a century earlier, her films reveal Pathé-Frères and Gaumont engaged in the making of *féeries* which exhibit the narrative conventions and scenic qualities of pantomime openings including fugitive lovers, pursuing parents and jealous lovers, and a benevolent agent who, with magical powers, transforms both characters and scenery. Final transformation and apotheosis scenes deploy a wealth of elaborate scenic effects, dancers, and exotically clothed supernumeraries. The splendours of Augustus Harris' and Oscar Barrett's spectacular pantomime final scenes are readily evoked, if not actually recreated. In French cinema, as in English pantomime, there was a ready market in the music halls and early cinemas for fairy stories depicting good and evil supernaturals who rescue and impede, vain and pompous rulers who use their authority capriciously, and innocent lovers reliant on magic to assist the continuation of their romance.

French *féerique* films reached British audiences in the first years of the twentieth century as Pathé, the world's most aggressive and powerful film producer, established marketing offices in London's Charing Cross Road. Pathé's weekly advertisements between 1905 to 1910 in the *Era* show that *féeries* and *diableries* were imported at an average of about one each fortnight, the number doubling in November and December in anticipation of sales to variety theatres for the Christmas holiday season. Pathé insisted that theatres buy and show entire films, not selected fragments. Typically, *The Haunted Grotto* (1909) sold at £6. 5s. 8d., but purchasers might have the film toned[19] for an added £1. 5s. *Miss Faust* (1909) sold at £11. 0s but was also available in a hand-coloured version[20] for a further £6. 6s.

From 1908, Pathé films increasingly replaced pantomime big-heads and trick scenery with time-lapse and stop-action effects. However, early French filmmakers also accepted that older theatrical technologies were suitable in achieving remarkable illusions, so it is common to find stage devices and film tricks utilised in the same sequence. *Féerique* films which are worthy of scrutiny for their proximity to Victorian pantomime include Méliès's *Éclipse du soleil en pleine lune,* Pathé's *Hallucinations d'un Pierrot,* Pathé's *Cochon danseur,* Gaumont's *L'Homme Aimanté,*[21] all from 1907. The film historian Tom Gunning adds to this list three Pathé films even closer in subject matter to English pantomime: *La Poule aux oeufs d'or* (1905), *Aladin ou la lampe merveilleuse* (1906),[22] and *Le Chat botté* (1903).[23] Monica Dall'Asta notes that the first female filmmaker, Alice Guy, made Solax's *Dick Whittington and His Cat* (1913) for British and American audiences.[24] Additionally, Tony Fletcher cites an uncatalogued *Robinson Crusoe,* shot in 1902 by G. A. Smith, that recreates an 1890s version at the Brighton Aquarium.[25]

However, because French films, apart from those by Méliès, are often inaccessible or viewable only under difficulty, theatre historians seeking to recover the narrative and technologies of late-Victorian pantomime in a readily accessible early twentieth-century French film may find that wish satisfied – or further stimulated – by Pathé's *Vie de Polichinelle* (1907), marketed in Britain and America as *Legend of Polchinella.*[26] Starring the French comedian Max Linder as Polichinelle (in English intertitles referred to as 'Harlequin' but, from his costume visibly Polichinelle or Punch), this brief film – in its use of a popular *commedia dell'arte* character, fugitive lovers, magical transformations and elaborate theatrical effects, the restoration and re-animation of a mutilated body, a supernatural benevolent agent, and a final splendid apotheosis – recapitulates numerous elements of pantomime openings and Harlequinades.

Vie de Polichinelle begins as Polichinelle, an automaton in a toyshop, is separated from his partner, a female automaton, when she is boxed and shipped to a ducal palace to amuse the noble's aristocratic guests. Despairing, Polichinelle is given a chance to recover this human-size doll when a fairy queen gives him life. When he absconds from the toyshop, riding a toy horse, the toymakers, angry that he has vanished, set off in pursuit. Their efforts to apprehend Polichinelle provide a sequence of scenes which use the machinery and illusions of the Victorian stage to discomfit the toymakers – leaving them hanging from the limbs of a tree which has suddenly risen from the stage floor – and facilitate Polichinelle's escape: a slide down a rocky precipice

Polichinelle, leading a platoon of torch-wielding sprites and elves (masked children), then arrives at a castle where the object of his love is entertaining a court of eighteenth century be-wigged aristocrats. To aid his abduction of the doll, the elves set small diversionary fires. As smoke rises from the salon floor, courtiers flee, deserting the automaton. The salon's walls – breakaway scenery – collapse, crushing and dismembering the automaton. Polichinelle enters and, undismayed, picks up the scattered pieces of the dismembered doll, wraps them into a bundle and carries his burden from the wreckage. Reaching the rocky chasm he had earlier traversed, Polichinelle creates by magic a rustic bridge, and crosses with his bundle into an elegant formal garden. There he re-animates the doll, by stop-action changing a mechanical toy into a human. Having eluded pursuers and acknowledging mutual love, Polichinelle and his fiancée enter a toyshop where life-size toys hang from the ceiling and walls. With a wave of his stick, Polichinelle restores these toys to life. After celebratory dances by the former toys, the back wall of the shop sinks into the floor to reveal a benevolent immortal and her train. This fairy queen, first transforming the shop into a palatial setting, oversees a sequence of increasingly lavish ballets culminating in an apotheosis in which a female soloist performs a 'serpentine dance' à la Löie Fuller.[27] The entire narrative has occupied a mere ten minutes.

Understandably, English films are the surest link to Victorian pantomime, but, with one single disappointing exception, these films are all Harlequinades intended for the holiday music-hall market. The odd film out is a brief forty-nine seconds of Herbert Campbell, best known for his pantomime partnerships with Dan Leno. A conspicuously large man here costumed as the child 'Little Bobby' for the 1899 Drury Lane pantomime *Cinderella,*[28] he is seated at a table, messily devouring what appear to be mashed potatoes from an oversized, overfilled bowl, then greedily downing a pint of beer before pulling faces at the camera. Bryony Dixon explains in a catalogue note that this footage was shot by 'British Biograph' (who also concurrently filmed Sir Herbert Beerbohm Tree in two scenes from Shakespeare's *King John*) as publicity for the Drury Lane pantomime and also used as a part of the entertainment at the Palace Theatre of Varieties in Shaftesbury Avenue.

Four films shot between 1912 and 1923 offer evidence of the continuity of the Harlequinade into the twentieth century. These films confirm that the Harlequinade was still essential to, or, at the very least, included in pantomime until the 1930s. They also reveal that the narrative dynamics – which brought the Regency Harlequinade into being and which sustained the mid-Victorian Harlequinade until pantomime became dominated by spectacle and music hall stars – had faded earlier in the twentieth century. There are no supernatural agents, virtuous or wicked, to conduct the initial transformation of the opening's characters into the archetypes of the Harlequinade. The Harlequinades either simply begin or require framing narratives which offer 'rational' everyday pretexts for Harlequinades taking place. All four films similarly confirm a Clown-centred Harlequinade but also point to the diminished, almost irrelevant, roles of Harlequin and Columbine. There is no pursuit by a Pantaloon avid to capture and separate the fugitive lovers, restoring Columbine to her original unsuitable suitor, often transformed into Clown. Clown seems altogether indifferent to the lovers' presence. There are no rivalries, no Clown assaults on Pantaloon. Harlequin still carries a bat, now elongated and curved, but makes only minor transformations to objects on the stage. In the main, he and Columbine appear as dancers, but there is little leaping through traps or flaps.

Thus, although Harlequin and Columbine are presences, they have no discernable functions within a dramatic narrative. Indeed, there is no narrative, apart from that of miscellaneous mischief: disrupting trade, exchanging babies in prams with live piglets, stealing sausages and geese, humiliating well-dressed citizens – but within that repertoire of mischief we witness many of pantomime's traditional sight-gags. Clown, Pantaloon, Policeman, and a cast of tradesmen and passers-by dominate activities. Because there is neither an overall narrative nor motive for pursuit, there is no call for a progression of Harlequinade scenes. Rather, all Harlequinade activity occurs in a single setting: a village street with shops. There are, however, framing episodes which use exterior settings.

The earliest Harlequinade film, *Harlequinade Let Loose*, filmed by the Hepworth Manufacturing Company in Walton-on-Thames[29] in 1912 or 1913, is such a mixture of indoor and outdoor footage. The framing pretext for this film is a drunken stage property-man's dream in which Harlequinade characters materialise and, predictably, create havoc in a street of shops, whitewashing customers in a barbershop, knocking over a pieman, ensnaring a billposter in his own papers and glue-buckets, and finally repairing to a tavern where, using the traps described in Moynet and the routines employed in haunted inn scenes, they disconcert the female tavernkeeper.

A rival film, *Here we Are Again* (1913),[30] similarly relies on a dream to frame the action. A policeman observes two thieves robbing a butcher and in a daydream transforms them into Clown and Pantaloon. A courting couple, spied in a park, are re-imagined as Columbine and Harlequin. The policeman's daydream brings all characters to the standard Harlequinade setting of a street in front of a pork butcher's shop where Harlequin transforms a donkey into Pantaloon and Clown. After stealing sausages and a ham, Clown falls into a mincing machine from which he emerges as a string of sausages. The film ends as the dreaming policeman, in his dream visited by Father Christmas, who brings him a gift of sergeant's stripes, is severely recalled to duty by his actual sergeant shaking him awake.

In the brief minute and a half of *Music Hall Act* (1914),[31] an incomplete Harlequinade film whose title is lost, Clown is implicated in a theft from a butcher's shop when Harlequin causes a crate labelled 'sausages' to appear in Clown's hands. A Policeman is drawn into the action as he attempts to apprehend Clown. The enraged butcher and other tradesmen join in a 'spill-and-pelt'[32] in which items of food – *papier maché* joints of meat and vegetables – are hurled at Clown and Pantaloon. Eventually, the butcher is lured into in his own mincing machine and emerges as a string of sausages.

The most complete Harlequinade, unimaginatively titled *Harlequinade,* was filmed nine years later in what was likely one of the Stoll theatres rather than a film studio. Stoll also drew from the casting of George Robey as Clown. Robey had graduated from an amateur performer to a professional comic and singer, first appearing in pantomime in 1891 and frequently thereafter, invariably as a dame in the opening and rarely continuing past the transformation scene into the Harlequinade. Although Robey made as many as twenty-five films, *Harlequinade* was his only appearance in film pantomimes.

Fortunately for the researcher, *Harlequinade* (1923)[33] exists in two versions, a completed film five minutes and fifty seconds in duration and an unedited reel of outtakes and 'rushes' which provides an additional thirty-three minutes of Harlequinade business, some tricks succeeding on the first take, others requiring re-takes and changes of camera position. Although there is a notional sequence of sight gags in the edited version, the rushes suggest that *Harlequinade*'s structure was wholly arbitrary, even to the point of filming multiple endings which include Pantaloon throwing sweets and Christmas crackers toward the camera[34], Father Christmas (Robey) visiting sleeping children and leaving gifts on their bed, a photo-montage of Harlequinade characters dancing on a chest containing either pies or sausages, and Clown and Pantaloon painting on the upstage wall a valedictory sign that eventually reads, 'A Happy New Year to You All'.

Other episodes include a lengthy narrative in which Pantaloon and Clown appropriate a free-standing machine erected to brush the clothing and polish the shoes of paying customers who pass by. They lure into their 'Wash and Brush-up' apparatus a burglar, whose loot they steal, a wealthy matron whose face they blacken with boot polish, whose necklaces they steal, and whose baby, in its pram, is exchanged with a butcher's piglet. An aristocratic gentleman is similarly assaulted. Eventually Clown and Pantaloon are pursued by a Policeman, who pockets all their swag, eluding pursuit by ducking into pivoting and double-hinged traps, whilst Harlequin causes the entire machine to vanish. As the brief police pursuit continues, Clown knocks over a pieman and robs a butcher's shop of sausages, employing the string of sausages as an escape rope, in the manner of the 'Indian rope trick', to ascend into the flies. Harlequin, however, makes the rope collapse.

As Robey's Harlequinade was exhibited in the bioscope programme during the 1923–24 Christmas season at each of Oswald Stoll's thirteen variety theatres, Robey himself was appearing in pantomime as Dame Trot in *Jack and the Beanstalk* at Glasgow's Alhambra Theatre. A photograph of Robey as Clown appeared in the *Era* heading an article[35] in which he expressed his wish to perform Clown in live pantomime but acknowledged that this wish was unlikely to be fulfilled. The Alhambra's Clown was Sam Foster.

In the past I have argued that earlier pantomimes, through their jokes, tricks-of-construction, and well-chosen topical scenery, offered a running index or a window onto their own specific eras. On the basis of its internal evidence, a well-constructed clever pantomime was immediately dateable, not only to the year but sometimes to the very month. At first glance and by contrast, these filmic Harlequinades, as much as the *féeries* and films of illusion, seem removed from such topicality. They seem bland and remote from any immediate preoccupations. There are no signs of war, no automobiles, no zepplins, no parodies of celebrities or noted eccentrics. Politics and politicians are remote. And yet in inter-class conflict, in the depiction of village life, and in pantomime's endless fascination with gadgets, there is a subdued topicality, albeit one of decades rather than years.

What is altogether missing from both Harlequinades and the continental and British films which offer simulacra of pantomime openings is any indication of music. Those Harlequinade films exhibited in larger urban music halls could call upon pit orchestras for accompaniment. Smaller picture-theatres had their resident pianists. In either case, the choice of music was left to the conductor or to the pianist. There were, consequently, no attempts to suggest topical songs and no evidence than any filmmaker imagined a song sheet, dropped from the flies or projected on the screen, to stimulate a communal sing-along. That lack of specified music may to some degree explain why Harlequin and

Columbine danced so little, or it may be that dance no longer held much appeal for cinema audiences.

Based on the, albeit limited, evidence of films that have survived, it is unlikely that there ever existed a film which depicted an entire Victorian pantomime with its obligatory opening, transformation scene in which the characters of the opening were transformed into those of the Harlequinade, followed by a Harlequinade with multiple changes of locales, the Harlequinade, in turn, followed by a 'dark' scene – the Harlequinade characters overtaken, captured, and Harlequin deprived of his magic bat, the pantomime concluding with a final scene of reconciliation, amity, and splendour. Such a film would have been too long and, as much to the point, not the material that English filmmakers elected to produce. Rather, some of the elements essential to a full Victorian pantomime were never filmed, and an approximation of full-length pantomime can only be pieced out from the several disparate sources I have described.

This lack of an entire pantomime, however, should not deter the researcher, the theatre scholar, who will come to recognise that in archival film are valuable clues to the intentions of the early moderns of France, America, and Britain to give cinema audiences the experiences and effects of theatrical pantomime and other analogous dramatic narratives: fairytales, magic, supernatural beings and events, illusions of gaudy splendour, and inexplicable changes which still relied on the machinery and trapwork of the stage as much, or sometimes more than, effects achieved by trick photography.

Additionally, the English Harlequinade films offer immediate connection to Victorian pantomime. These brief films, some of them incomplete, each lacking a narrative of pursuit and capture foiled by tricks and trapwork do necessarily await close study. Harlequinade films were made because audiences still expected the presence of Clown, Pantaloon, Policeman, Columbine, and Harlequin in their holiday entertainment, even if their appearance was brief and, at times, more *pro forma* than spirited. These audiences tolerated, even enjoyed, the same sight-gags and hackneyed routines that amused their Victorian ancestors. Victorian pantomime on twentieth-century film is not an anomaly or paradox but evidence of the survival of a species of entertainment that outlived one era and that – briefly – had potency on the cusp of a new age. As much to the point, Victorian pantomime on twentieth-century film is a tangible research resource awaiting close examination and intensive analysis. In this chapter I have merely offered a rough map to this resource, sketching in the most rudimentary description of trails and topography. However, as Umberto Eco and other writers have reminded us, 'a map is not the territory'.[36] The investigator has only to mount her or his own expedition.

Notes

1. Archives holding promising sources include 1) British Film Institute National Film and Television Archive, 2) Cinémathèque Française / Bibliothèque du film, 3) Motion Picture, Broadcasting and Recorded Sound Division, (US) Library of Congress, 4) Film Department, Museum of Modern Art, 5) Cinémathèque Royale de Belgique, 6) Deutsche Kinemathek, 7) Département des Arts du Spectacle de la Bibliothèque Nationale de France, 8) American Film Academy, 9) George Eastman House. 10) Gaumont Pathé Archives.
2. Early film was sold to exhibitors by the linear foot. Thus, although a complete narrative may have been of a specified length, exhibitors frequently bought only preferred portions. Surviving films reflect these sales and purchasing practices. *Vie de Polichinelle* is such a film, surviving in two versions, one with its harlequinade, one without this footage.
3. Bryony Dixon, *Chaplin in Context Project: A Catalogue of Music Hall Related Films 1895–1930 held by the National Film and Television Archive* (London: British Film Institute, 2003). Available on-line: chaplin.bfi.org.uk/resources/bfi/pdf/chaplin-in-context.pdf
4. Additionally, there has been a miscellany of discoveries and clarifications made by Tony Fletcher, a researcher-archivist at London's Cinema Museum. His findings have not enjoyed the public scrutiny accorded to those of Dixon, Lewinsky, and Solomon, but Fletcher is immediately accessible to researchers and generous in supplying information. The Cinema Museum, The Master's House, 2 Dugard Way, London SE11 4TH; Tel.: +44 (0)20 7840 2200; Fax: +44 (0)20 7840 2299; email: info@cinemamuseum.org.uk
5. Charles Musser, *Edison Motion Pictures, 1890–1900: An Annotated Filmography* (Washington, DC: Smithsonian Institution Press & Pordenone: Le Giornate del Cinema Muto, 1997).
6. Kemp R. Niver, *Early Motion Pictures: The Paper Print Collection in the Library of Congress*, ed. Bebe Bergsten (Washington: Library of Congress, 1985), p. 327.
7. Paolo Cherchi-Usai, 'Film Measurement Tables', *Burning Passions: An Introduction to the Study of Silent Cinema,* trans. Emma Sansone Rittle (London: British Film Institute, 1994), p. 93. All running times given in this essay are derived from these tables or from actual timings of films viewed at 16 or 18 fps.
8. Georges Moynet, 'Décor à Trucs avec Trappes Anglaises', *Trucs et Décors: La Machinerie Théatrale (*Paris: la Librairie Illustrée, 1893), pp. 101–21.
9. Many early films survive in the US Library of Congress' 'Paper Print' Collection. In 1894, uncertain how to classify motion pictures but willing to grant international copyright to filmmakers, the Library of Congress stipulated that filmmakers seeking copyright were to deposit their films as continuous positive images printed onto narrow paper strips similar to the rolls in cash registers or other tabulating instruments. Some of these unprojectable

strips were but a few feet long. Others ran for many thousands of feet. This unusual deposit system continued until 1912, when actual movie films replaced opaque prints. These paper films have now been transferred to conventional film stock and are viewable on application. See Kemp Niver's catalogue (above).

10. Georges Méliès, *Faust aux Infers,* Montreuil, autumn, 1903, released as *The Damnation of Faust,* Library of Congress Paper Print Film FLA5033.
11. David Robinson, *Georges Méliès, Father of Film Fantasy* (London: British Film Institute, 1993), pp. 5 & 6.
12. Elizabeth Ezra, *Georges Méliès,* French Film Directors series (Manchester & New York: Manchester University Press, 2000), p. 9.
13. David Mayer, *Harlequin in His Element: English Pantomime, 1806–1836* (Cambridge, MA: Harvard University Press, 1969), pp. 130–6.
14. Although, later in this essay, I describe a 'sink' effect in *Vie de Polichinelle,* and the effect of a wall of boulders rising, as if by magic, in a field is used in Ambrosio's *Topolini Riconocenti* (1908).
15. This DVD compilation, which includes nearly all of his surviving films and fragments, many gathered by Serge Bromberg and Lobster Films, was released in 2008 by Flicker Alley as *Georges Méliès, First Wizard of the Cinema (1896–1913).* http://flickeralley.com/fa_melies_01.html
16. Matthew Solomon, 'Magic in Film', *Catologo, Le Giornate del Cinema Muto 2006/ 25*th Pordenone Silent Film Festival (Pordenone: Le Giornate del Cinema Muto, 2006), pp. 97–106.
17. These Paul films are included in the 2006 BFI DVD *R. W. Paul: The Collected Films, 1895–1908.* http://filmstore.bfi.org.uk/acatalog/info_2786.html
18. Mariann Lewinsky, '100 Years Ago: The Films of 1907' *Catologo, Il Cinema Ritrovato,* 21o Edizione (Bologna: Cinetecca del Comune di Bologna, 2007), pp. 64–80.
19. Toning was a chemical process which intensified and added a single colour to a black-and-white film.
20. Films were brilliantly and expertly coloured with a stenciling process. It is uncertain how many exhibitors bought films which had been coloured, but a surprising number of such films survive, perhaps because these were more expensive and therefore deemed more worthy of special care.
21. This film and other Gaumont féerique films have been issued on a new 7-disc DVD set, *Gaumont, Le Cinéma Premier, 1897–1913,* vol. 1: *Alice Guy, Louis Feuillade, Léonce Perret,* Paris, 2008.
22. This film is available on video and DVD in Vol. 3, *Experimentation and Discovery* in Kino Video and the British Film Institute's multi-volume set, *The Movies Begin.*
23. Tom Gunning, '"Now You See It, Now You Don't": The Temporality of the Cinema of Attractions', *Silent Film,* ed. Richard Abel (New Brunswick, NJ: Rutgers University Press, 1996), p. 81.
24. Monica Dall'Asta, *Alice Guy, Memorie di una Pioniera del Cinema* (Bologna: Cineteca di Bologna, 2008), p. 205.
25. Private correspondence, October 2007. Laura Bayley took the role of Crusoe. The whereabouts or condition of this film are not disclosed.
26. This film, directed by Albert Capellani and sold by Pathé as a *féerie,* exists in two versions, neither wholly complete in itself: a print, 410 metres long

with Russian intertitles, owned by the BFI, and on a DVD *The Comedy of Max Linder*, an anthology of fourteen films shot between 1905 and 1913. Phoenix AZ: Grapevine Video, 2003. www.grapevinevideo.com

27. Because performers were rarely identified, the serpentine dancer may actually be Löie Fuller. Pathé filmed her performing this dance in 1901 and again in 1905 and may have, through optical processing, included actual performance footage in their apotheosis.
28. 'Herbert Campbell as "Little Bobby" (1899)', 49 ft., British Mutoscope and Biograph Company, Bryony Dixon, *Chaplin in Context Project*, p. 4.
29. *Harlequinade Let Loose* (1912 or 1913), 478 ft. (approximately 7′ 30″), Hepworth Manufacturing Company. Dixon's catalogue gives the date 1912. Whimsical Walker's autobiography, *From Sawdust to Windsor Castle* (London: Stanley Paul, 1922), pp. 236–8, states that it was filmed in 1913. Walker, working under the name of James Darling, made additional films for Hepworth.
30. *Here We Are Again* (1913), 489 ft, (approximately 8′) produced either by Clarendon Films or George Cricks.
31. *Music Hall Act* (1914) 90 ft. (approximately 1′ 30″) (its actual title or producer unknown, but possibly the work of Bamforths, the postcard manufacturers, of Holmfirth, Yorks.).
32. The 'spill and pelt', where quantities of *papier maché* and cloth property meats, vegetables, pies and beverages were thrown about the stage by the Harlequinade characters (an ancestral form of the more modern 'slosh' or 'decorators' scene), was already extant in the 1840s and even then derided by critics as insufficiently amusing to merit staging. A spill and pelt is depicted as a 'pantomime storm' in 'Alfred Crowquill's' 1849 strip cartoon *Pantomime as it Was, Is, and Will Be*. The spill and pelt in *Here We Are Again* is the only example of this still-traditional episode recorded on film.
33. *Harlequinade* (1923) completed version 356 ft. (approximately 5′ 50″); reel of 'rushes' and outtakes 2000 ft. (approximately 33′). Stoll Films.
34. Claud Zola, an acrobat who appeared in (London) Lyceum pantomimes as a 'skin man', i.e., an actor who impersonated such animals as Dick Whittington's cat and Jack (and the Beanstalk)'s cow, was annually assigned to stage the Harlequinade and to undertake the role of Clown, not performing the Harlequinade until a full week after the pantomime had opened. A part of Clown's duties was to throw candies and crackers, weighted so that they might fly further into the auditorium, at the pantomime's end. With this assignment for the last Lyceum pantomime, *The Queen of Hearts* (1939), Zola was confronted at a matinee with the then-Princesses Elizabeth and Margaret leaning from their auditorium box and pleading for sweets and crackers. Although Zola had been warned by the theatre's management not to hurl anything at the Royal Box, the two children were leaning so precariously that, rather than risk their falling, he threw a heavy cracker. As Zola told me in 1968, 'It hit the Queen Mum right on the nut'.
35. *Era*, 2 January 1924, p. 21.
36. Umberto Eco, epigraph to Chapter 83, quoting Alfred Korzybski, *Foucault's Pendulum*, (London: Picador Books, 1990), p. 456.

Select Bibliography

Adams, W. Davenport. 'The Decline of Pantomime'. In *The Theatre n.s V* (1 February, 1882).

Anon. *Revelations of Life in Nottingham by the English Asmodeus*. Nottingham: C. B Truman, 'Telegraph' Office [1860].

Archer, William. *The Theatrical 'World' of 1893*. London: Walter Scott, 1894.

Bailey, Peter. *Leisure and Class in Victorian England: Rational Recreation and the Contest for Control 1830–1885*. London: Methuen, reprinted 1987.

Baker, Roger. *Drag*. London: Cassell, 1994.

Barker, Kathleen M. D. 'The Performing Arts in Five Provincial Towns, 1840–1870', unpublished doctoral thesis, University of Leicester, 1982.

Beerbohm, Max. *More Theatres 1898–1903*, ed. Rupert Hart-Davies. London: Rupert Hart-Davies, 1969.

Beerbohm, Max. *Last Theatres 1904–1910*. London: Rupert Hart-Davis, 1970.

Booth, Michael R., ed. *English Plays of the Nineteenth Century: V. Pantomimes, Extravaganzas and Burlesques*. Oxford: Oxford University Press, 1976.

Booth, Michael R. *Victorian Spectacular Theatre*. London: Routledge & Kegan Paul, 1981.

Booth, Michael R. *Theatre in the Victorian Age*. Cambridge: Cambridge University Press, 1991.

Brandreth, Gyles. 'Oh yes it is!' *Sunday Telegraph Review* 10 December, 2000.

Bratton, Jacky. *New Readings in Theatre History*. Cambridge: Cambridge University Press, 2003.

Bratton, Jacky and Ann Featherstone. *The Victorian Clown*. Cambridge: Cambridge University Press, 2006.

Burd, Van Akin. *Christmas Story: John Ruskin's Venetian Letters of 1876–1877*. University of Delaware Press, 1991.

Burnand, F. C. *Records and Reminiscences: Personal and General*, 2 vols. London: Methuen, 1904.

Butler, Judith. *Gender Trouble: Feminism and the Subversion of Identity*. New York and London: Routledge, 1990.

Byron, H. J. *Aladdin* ed. Gyles Brandreth. London: Davis Poynter, 1971.

Byron, H. J. *Cinderella* ed. Gyles Brandreth. London: Davis Poynter, 1971.

Byron, H. J. *Plays by H. J. Byron* ed. Jim Davis. Cambridge: Cambridge University Press, 1984.

Carlisle, Janice. 'Spectacle as Government, Dickens and the Working-Class Audience'. In Sue-Ellen Case and Janelle Reinelt, eds., *The Performance of Power: Theatrical Discourse and Politics*. Iowa City: University of Iowa Press, 1991.

Clinton-Baddeley, V. C. *Some Pantomime Pedigrees*. London: The Society for Theatre Research, 1963.

Cohen, Morton N., ed., with the assistance of Roger Lancelyn Green. *The Letters of Lewis Carroll*, 2 vols. London: Macmillan, 1979.

Coleman, John. *Players and Playwrights I Have Known*, 2 vols. London: Chatto and Windus, 1888.

Cook, E. T. and Alexander Wedderburn, eds. *The Complete Works of John Ruskin*, 39 vols. London: George Allen, 1903–1912.

Davis, Jim, ed. *The Britannia Diaries 1863–1875*. London: The Society for Theatre Research, 1992.

Davis, Jim. 'Imperial Transgressions: the Ideology of Drury Lane Pantomime in the Late Nineteenth Century'. In *New Theatre Quarterly* 12.46 (May 1996).

Davis, Jim. 'Boxing Day'. In Tracy C. Davis and Peter Holland, eds. *The Performing Century: Nineteenth Century Theatre's History*. Basingstoke: Palgrave Macmillan, 2007.

Davis, Jim and Victor Emeljanow. *Reflecting the Audience: London Theatregoing. 1840–1880*. Hatfield: University of Hertfordshire Press, 2001.

Davis, Tracy C. 'The Employment of Children in the Victorian Theatre'. In *New Theatre Quarterly* 2.6 (May, 1986).

Davis, Tracy C. *The Economics of the British Stage 1800–1914*. Cambridge: Cambridge University Press, 2000.

Davis, Tracy C. 'What are Fairies For?' In Tracy C. Davis and Peter Holland eds., *The Performing Century*. Basingstoke: Palgrave Macmillan, 2007.

Dickens, Charles. 'Holding Up the Mirror'. In *All the Year Round* (29 September, 1860).

Dickens, Charles. 'The Uncommercial Traveller'. In *All the Year Round* (25 February 1860).

Dickens Junior, Charles. 'On the Decline of Pantomime'. In *The Theatre* vol. 27 (January 1896).

Dixon, Bryony. *Chaplin in Context Project: A Catalogue of Music Hall Related Films 1895–1930 held by the National Film and Television Archive*. London: British Film Institute, 2003.

Ezra, Elizabeth. *Georges Méliès*, French Film Directors series. Manchester & New York: Manchester University Press, 2000.

Findlater, Richard, ed. *Memoirs of Joseph Grimaldi* by Charles Dickens. London, 1838; revised edition London: MacGibbon & Kee Ltd., 1968.

Fleetwood, Frances. *Conquest: The Story of a Theatre Family*. London: W. H. Allen, 1953.

Frow, Gerald. *'Oh, Yes It Is': A History of Pantomime*. London: BBC Books, 1985.

Foulkes, Richard. *Lewis Carroll and the Victorian Stage: Theatricals in a Quiet Life*. Aldershot: Ashgate, 2005.

Gillies, Midge. *Marie Lloyd: The One and Only*. London: Orion, 1999.

Glover, J. M. *Jimmy Glover His Book*. London: Methuen, 1911.

Harris, Augustus. 'Spectacle'. In *The Magazine of Art* vol. 12, 1889.

Hibbert, H. G. A *Playgoer's Memories*. London: Grant Richards, 1920.

Hickory Wood, J. *Dan Leno*. London: Methuen, 1905.

Holland, Peter. 'The Play of Eros: The Paradoxes of Gender in English Pantomime'. In *New Theatre Quarterly* 13.51 (Aug., 1997).

Jackson, Russell, ed. *Victorian Theatre*. London: A & C Black, 1989.

Lefebvre, Henri. *The Production of Space* (1974), trans., Donald Nicholson-Smith. Oxford: Blackwell, 1991.

Lemon, Mark. 'An Actor's Holiday'. *London Society*, vol. XII (The Christmas number for 1867).

Levy, E. Lawrence. *Birmingham Theatrical Reminiscences: Jubilee Recollections (1870–1920)*. Birmingham: Messrs. J. G. Hammond & Co. Ltd. [1920].

Mander, Raymond and Joe Mitchenson. *Pantomime: A Story in Pictures*. London: Peter Davies, 1973.
Mayer, David. *Harlequin in his Element: The English Pantomime 1806–1836*. Cambridge, Mass: Harvard University Press, 1969.
Mayer, David Mayer. 'The Sexuality of Pantomime'. In *Theatre Quarterly* 4.13 (March/April, 1974).
Moody, Jane. *Illegitimate Theatre in London, 1770–1840*. Cambridge: Cambridge University Press, 2000.
Moynet, Georges. *Trucs et Décors: La Machinerie Théatrale*. Paris: la Librairie Illustrée, 1893.
Niver, Kemp R. *Early Motion Pictures: The Paper Print Collection in the Library of Congress*, ed. Bebe Bergsten. Washington: Library of Congress, 1985.
O'Brien, John. *Harlequin Britain: Pantomime and Entertainment*. Baltimore and London: The Johns Hopkins University Press, 2004.
Planché, J. R. *Recollections and Reflections*, 2 vols. London: Tinsley Brothers, 1872.
Planché, J. R. *Plays by J. R. Planché*, edited by Donald Roy. Cambridge: Cambridge University Press, 1986.
Reed, Langford. *The Life of Lewis Carroll*. London: W. & G. Foyle Ltd., 1932.
Robbins, Norman. *Slapstick and Sausages: The Evolution of British Pantomime*. Tiverton: Trapdoor Publications, 2002.
Robinson, David. *Georges Méliès, Father of Film Fantasy*. London: British Film Institute, 1993.
Robinson, Jo. 'Mapping performance culture: locating the spectator in theatre history'. In *Nineteenth Century Theatre and Film*, 31.1 (2004).
Sala, G. A. 'Pantomime and Pandemonium: Two Nights in the New Cut'. In *Belgravia*, vol. XI (April 1870).
Salberg, Derek. *Once Upon a Pantomime*. Luton: Cortney Publications, 1981.
Schiach, Morag. *Discourse on Popular Culture: Class, Gender and History in Cultural Analysis 1730 to the Present*. Cambridge: Polity Press; and Stanford: Stanford University Press, 1989.
Scott, Clement. *The Drama of Yesterday and Today*, 2 vols. London: Macmillan and Co., 1899.
Scott, Clement and Cecil Howard. *The Life and Reminiscences of E. L. Blanchard*, 2 vols. London: Hutchinson & Co., 1891.
Senelick, Laurence. *The Changing Room: Sex, Drag and the Theatre*. London: Routledge, 2000.
Shaw, G. B. *Our Theatres in the Nineties*, 3 vols. London: Constable, 1954.
Shepard, Ernest. *Drawn from Memory*. London: Methuen, 1971.
Stirling, Edward. *Old Drury Lane*, 2 vols. London: Chatto and Windus, 1881.
Sullivan, Jill A. 'The Business of Pantomime: Regional Productions 1865–1892'. Unpublished doctoral thesis, University of Nottingham, 2005.
Sullivan, Jill A. 'Victorian Pantomime Libretti and the Reading Audience'. In Graham Allen, Siobhán Collins, Carrie Griffin and Mary O'Connell, eds., *Making Books, Shaping Readers* (forthcoming, Pickering & Chatto, 2010).
Taylor, Millie. *British Pantomime Performance*. Bristol: Intellect Books, 2007.
Thackeray, W. M. 'Round About the Christmas-Tree'. *Roundabout Papers, reprinted from the Cornhill Magazine*. New York: Harper and Brothers Publishers, 1863.
Turnbull, Olivia. *Bringing Down the House*. Bristol and Chicago: Intellect, 2008.

Varty, Anne. *Children and Theatre in Victorian Britain: 'All Work, No Play'*. Basingstoke: Palgrave Macmillan, 2008.

Wagner, Leopold. *The Pantomimes and All About Them*. London: John Heywood, 1881.

Wakeling, Edward, ed., *Lewis Carroll's Diaries*, 10 vols. Luton: the Lewis Carroll Society, 1999.

Walch, Garnet. *Harlequin Felix or Harlequin Laughing Jackass and the Magic Bat*, ed. Veronica Kelly. St. Lucia: University of Queensland Press, 1988.

Walpole, Hugh. *Jeremy*. London: Cassell, 1919.

Weltman, Sharon. *Performing the Victorian: John Ruskin and Identity in Theater, Science, and Education*. Columbus: Ohio State University Press, 2007.

Williams, Margaret. *Australia on the Popular Stage 1829–1929: An Historical Entertainment in Six Acts*. Melbourne: Oxford University Press, 1983.

Wilson, A. E. *Christmas Pantomime: The Story of a British Institution*. London: Allen & Unwin, 1934.

Wilson, A. E. *Pantomime Pageant*. London: Stanley Paul & Co Ltd., 1946.

Worrall, David. *Theatric Revolution: Drama, Censorship and Romantic Period Subcultures 1773*–1832. Oxford: Oxford University Press, 2006.

Worrall, David. *Harlequin Empire: Race, Ethnicity and the Drama of the Popular Enlightenment*. London: Pickering and Chatto (Publishers) Limited, 2007.

Newspapers and periodicals

All the Year Round

Belgravia

Birmingham Daily Gazette

Birmingham Daily Mail

Birmingham Daily Post

Birmingham Journal

Chambers's Journal of Popular Literature, Science and Art

Daily News

Daily Telegraph

Era

Era Almanack

Examiner

Illustrated London News

London Society

Nottingham Journal

Nottingham and Midland Counties Daily Express

Nottingham Review

Nottinghamshire Guardian

Pall Mall Gazette

Punch

Reynolds's News

Saturday Review
Sketch
The Sphinx
Star
The Theatre
The Times

Index